C++ DESIGN PATTERNS AND DERIVATIVES PRICING

Mathematics, Finance and Risk

Editorial Board

C++ DESIGN PATTERNS AND DERIVATIVES PRICING

MARK S. JOSHI
Royal Bank of Scotland Group

CAMBRIDGE
UNIVERSITY PRESS

PUBLISHED BY THE PRESS SYNDICATE OF THE UNIVERSITY OF CAMBRIDGE
The Pitt Building, Trumpington Street, Cambridge, United Kingdom

CAMBRIDGE UNIVERSITY PRESS
The Edinburgh Building, Cambridge CB2 2RU, UK
40 West 20th Street, New York, NY 10011–4211, USA
477 Williamstown Road, Port Melbourne, VIC 3207, Australia
Ruiz de Alarcón 13, 28014 Madrid, Spain
Dock House, The Waterfront, Cape Town 8001, South Africa

http://www.cambridge.org

First published 2004
Reprinted with corrections 2005, 2006

Printed in the United Kingdom at the University Press, Cambridge

Typeface Times 11/14 pt.   *System* LATEX 2$_\varepsilon$ [TB]

A catalogue record for this book is available from the British Library

Library of Congress Cataloguing in Publication data available

Joshi, M.S. (Mark Suresh), 1969–
C++ design patterns and derivatives pricing / M.S. Joshi.
p.   cm. – (Mathematics, finance, and risk)
Includes bibliographical references and index.
ISBN 0 521 83235 7
1. Derivative securities–Prices–Mathematical models.   2. C++ (Computer program language)
3. Business mathematics.   I. Title.   II. Series.
$'$57 HG6024.A3J665 2004
332.64 $'$02855133 – dc22   2003065293

ISBN 0 521 83235 7 hardback

To Jane

Contents

Preface

This book is aimed at a reader who has studied an introductory book on mathematical finance and an introductory book on C++ but does not know how to put the two together. My objective is to teach the reader not just how to implement models in C++ but more importantly how to think in an object oriented way. There are already many books on object-oriented programming, however, the examples tend not to feel real to the financial mathematician so in this book we work exclusively with examples from derivatives pricing.

We do not attempt to cover all sorts of financial models but instead examine a few in depth with the objective at all times of using them to illustrate certain OO ideas. We proceed largely by example and our design process is not one to be recommended because we rewrite our designs as new concepts are introduced, instead of working out a great design at the start. Whilst this approach is not optimal from a design standpoint, it is more pedagogically accessible. An aspect of this is that our examples are designed to emphasize design principles rather than to illustrate other features of coding such as numerical efficiency or exception safety.

We commence by introducing a simple Monte Carlo model which does not use OO techniques but rather is the simplest procedural model for a pricing a call option one could write. We examine its shortcomings and discuss how classes naturally arise from the concepts involved in its construction.

In Chapter 2, we move on to the concept of encapsulation – the idea that a class allows to express a real-world analogue and its behaviours precisely. In order to illustrate encapsulation, we look at how a class to define the pay-off of a vanilla option could be defined. We also see that the class we have defined has certain defects, and this naturally leads on to the open-closed principle.

In Chapter 3, we see how a better pay-off class can be defined by using inheritance and virtual functions. This raises technical issues involving destruction and passing arguments which we address. We also see how this approach is compatible with the open–closed principle.

Using virtual functions causes problems regarding the copying of objects of unknown type, and in Chapter 4 we address these problems. We do so by introducing virtual constructors and the bridge pattern. We digress to discuss the 'rule of three' and the slowness of `new`. The ideas are illustrated via a vanilla options class and a parameters class.

With these new techniques at our disposal, we move on to looking at more complicated design patterns in Chapter 5. We first introduce the strategy pattern that express the idea that decisions on part of an algorithm can be deferred by delegating responsibilities to an auxiliary class. We then look at how templates can be used to write a wrapper class that removes a lot of our difficulties with memory handling. As an application of these techniques, we develop a convergence table using the decorator pattern.

In Chapter 6, we look at how to develop a random numbers class. We first examine why we need a class and then develop a simple implementation which provides a reusable interface and an adequate random number generator. We use the implementation to introduce and illustrate the adapter pattern, and to examine further the decorator pattern.

We move on to our first non-trivial application in Chapter 7, where we use the classes developed so far in the implementation of a Monte Carlo pricer for path-dependent exotic derivatives. As part of this design, we introduce and use the template pattern. We finish with the pricing of Asian options.

We shift from Monte Carlo to trees in Chapter 8. We see the similarities and differences between the two techniques, and implement a reusable design. As part of the design, we reuse some of the classes developed earlier for Monte Carlo.

We return to the topic of templates in Chapter 9. We illustrate their use by designing reusable solver classes. These classes are then used to define implied volatility functions. En route, we look at function objects and pointers to member functions. We finish with a discussion of the pros and cons of templatization.

In Chapter 10, we look at our most advanced topic: the factory pattern. This patterns allows the addition of new functionality to a program without changing any existing files. As part of the design, we introduce the singleton pattern.

Our final chapter reviews, assesses and organizes the patterns we have covered. In particular, we classify them as creational, structural and behavioural patterns. We finish with a list of further books to read.

All the code in this book is included on the accompanying CD. The book's website is

`www.markjoshi.com/design`

and any bugfixes will be posted there.

Acknowledgements

I am grateful to the Royal Bank of Scotland for providing a stimulating environment in which to learn, study and do mathematical finance. Most of my views on coding C++ and financial modelling have been developed during my time working there. My understanding of the topic has been formed through daily discussions with current and former colleagues including Chris Hunter, Peter Jäckel, Dherminder Kainth, Sukhdeep Mahal, Robin Nicholson and Jochen Theis. I am also grateful to a host of people for their many comments on the manuscript, including Alex Barnard, Dherminder Kainth, Rob Kitching, Sukhdeep Mahal, Nadim Mahassen, Hugh McBride, Alan Stacey and Patrik Sundberg. I would also like to thank David Tranah and the rest of the team at Cambridge University Press for their careful work and attention to detail. Finally my wife has been very supportive.

1

A simple Monte Carlo model

1.1 Introduction

In the first part of this book, we shall study the pricing of derivatives using Monte Carlo simulation. We do this not to study the intricacies of Monte Carlo but because it provides many convenient examples of concepts that can be abstracted. We proceed by example, that is we first give a simple program, discuss its good points, its shortcomings, various ways round them and then move on to a new example. We carry out this procedure repeatedly and eventually end up with a fancy program. We begin with a routine to price vanilla call options by Monte Carlo.

1.2 The theory

We commence by discussing the theory. The model for stock price evolution is

$$dS_t = \mu S_t dt + \sigma S_t dW_t, \tag{1.1}$$

and a riskless bond, B, grows at a continuously compounding rate r. The Black–Scholes pricing theory then tells us that the price of a vanilla option, with expiry T and pay-off f, is equal to

$$e^{-rT} \mathbb{E}(f(S_T)),$$

where the expectation is taken under the associated risk-neutral process,

$$dS_t = r S_t dt + \sigma S_t dW_t. \tag{1.2}$$

We solve equation (1.2) by passing to the log and using Ito's lemma; we compute

$$d \log S_t = \left(r - \frac{1}{2}\sigma^2 \right) dt + \sigma dW_t. \tag{1.3}$$

As this process is constant-coefficient, it has the solution

$$\log S_t = \log S_0 + \left(r - \frac{1}{2}\sigma^2 \right) t + \sigma W_t. \tag{1.4}$$

1

Since W_t is a Brownian motion, W_T is distributed as a Gaussian with mean zero and variance T, so we can write

$$W_T = \sqrt{T} N(0, 1), \tag{1.5}$$

and hence

$$\log S_T = \log S_0 + \left(r - \frac{1}{2}\sigma^2 \right) T + \sigma \sqrt{T} N(0, 1), \tag{1.6}$$

or equivalently,

$$S_T = S_0 e^{(r - \frac{1}{2}\sigma^2)T + \sigma \sqrt{T} N(0,1)}. \tag{1.7}$$

The price of a vanilla option is therefore equal to

$$e^{-rT} \mathbb{E}\left(f\left(S_0 e^{(r - \frac{1}{2}\sigma^2)T + \sigma \sqrt{T} N(0,1)} \right) \right).$$

The objective of our Monte Carlo simulation is to approximate this expectation by using the law of large numbers, which tells us that if Y_j are a sequence of identically distributed independent random variables then with probability 1 the sequence

$$\frac{1}{N} \sum_{j=1}^{N} Y_j$$

converges to $\mathbb{E}(Y_1)$.

So the algorithm to price a call option by Monte Carlo is clear. We draw a random variable, x, from an $N(0, 1)$ distribution and compute

$$f\left(S_0 e^{(r - \frac{1}{2}\sigma^2)T + \sigma \sqrt{T} x} \right)$$

where $f(S) = (S - K)_+$. We do this many times and take the average. We then multiply this average by e^{-rT} and we are done.

1.3 A simple implementation of a Monte Carlo call option pricer

A first implementation is given in the program `SimpleMCMain1.cpp`.

Listing 1.1 (`SimpleMCMain1.cpp`)

```
//  requires Random1.cpp


#include <Random1.h>
#include <iostream>
#include <cmath>
using namespace std;
```

```cpp
double SimpleMonteCarlo1(double Expiry,
                        double Strike,
                        double Spot,
                        double Vol,
                        double r,
                        unsigned long NumberOfPaths)
{
    double variance = Vol*Vol*Expiry;
    double rootVariance = sqrt(variance);
    double itoCorrection = -0.5*variance;

    double movedSpot = Spot*exp(r*Expiry +itoCorrection);
    double thisSpot;
    double runningSum=0;

    for (unsigned long i=0; i < NumberOfPaths; i++)
    {
    double thisGaussian = GetOneGaussianByBoxMuller();
    thisSpot = movedSpot*exp( rootVariance*thisGaussian);
    double thisPayoff = thisSpot - Strike;
    thisPayoff = thisPayoff >0 ? thisPayoff : 0;
    runningSum += thisPayoff;
    }

    double mean = runningSum / NumberOfPaths;
    mean *= exp(-r*Expiry);
    return mean;
}

int main()
{
    double Expiry;
    double Strike;
    double Spot;
    double Vol;
    double r;
    unsigned long NumberOfPaths;
    cout << "\nEnter expiry\n";
    cin >> Expiry;
```

```
    cout << "\nEnter strike\n";
    cin >> Strike;

    cout << "\nEnter spot\n";
    cin >> Spot;

    cout << "\nEnter vol\n";
    cin >> Vol;

    cout << "\nr\n";
    cin >> r;

    cout << "\nNumber of paths\n";
    cin >> NumberOfPaths;

    double result = SimpleMonteCarlo1(Expiry,
                                      Strike,
                                      Spot,
                                      Vol,
                                      r,
                                      NumberOfPaths);

    cout <<"the price is " << result << "\n";

    double tmp;
    cin >> tmp;

 return 0;
}
```

Our program uses the auxiliary files Random1.h and Random1.cpp.

Listing 1.2 (Random1.h)

```
#ifndef RANDOM1_H
#define RANDOM1_H

double GetOneGaussianBySummation();
double GetOneGaussianByBoxMuller();
#endif
```

Listing 1.3 (Random1.cpp)

```cpp
#include <Random1.h>
#include <cstdlib>
#include <cmath>

// the basic math functions should be in namespace
// std but aren't in VCPP6
#if !defined(_MSC_VER)
using namespace std;
#endif

double GetOneGaussianBySummation()
{
double result=0;

for (unsigned long j=0; j < 12; j++)
result += rand()/static_cast<double>(RAND_MAX);

result -= 6.0;

return result;
}

double GetOneGaussianByBoxMuller()
{
double result;

double x;
double y;

double sizeSquared;
do
{
x = 2.0*rand()/static_cast<double>(RAND_MAX)-1;
y = 2.0*rand()/static_cast<double>(RAND_MAX)-1;
sizeSquared = x*x + y*y;
}
while
( sizeSquared >= 1.0);
```

```
result = x*sqrt(-2*log(sizeSquared)/sizeSquared);

return result;
}
```

We first include the header file Random1.h. Note that the program has <Random1.h> rather than "Random1.h". This means that we have set our compiler settings to look for header files in the directory where Random1.h is. In this case, this is in the directory C/include. (In Visual C++, the directories for include files can be changed via the menus tools, options, directories.)

Random1.h tells the main file that the functions

```
double GetOneGaussianBySummation()
```

and

```
double GetOneGaussianByBoxMuller()
```

exist. We include the system file iostream as we want to use cin and cout for the user interface. The system file cmath is included as it contains the basic mathematical functions exp and sqrt.

We have the command using namespace std because all the standard library commands are contained in the namespace std. If we did not give the using directive, then we would have to prefix all their uses by std::, so then it would be std::cout rather than cout.

The function SimpleMonteCarlo1 does all the work. It takes in all the standard inputs for the Black–Scholes model, the expiry and strike of the option, and in addition the number of paths to be used in the Monte Carlo.

Before starting the Monte Carlo we precompute as much as possible. Thus we compute the variance of the log of the stock over the option's life, the adjustment term $-\frac{1}{2}\sigma^2 T$ for the drift of the log, and the square root of the variance. Whilst we cannot precompute the final value of spot, we precompute what we can and put it in the variable movedSpot.

We initialize the variable, runningSum, to zero as it will store the sum so far of the option pay-offs at all times.

We now loop over all the paths. For each path, we first draw the random number from the $N(0, 1)$ distribution using the Box–Muller algorithm and put it in the variable thisGaussian.

The spot at the end of the path is then computed and placed in thisSpot. Note that although our derivation of the SDE involved working with the log of the spot, we have carefully avoided using log in this routine. The reason is that log and exp are slow to compute in comparison to addition and multiplication, we therefore want to make as few calls to them as possible.

We then compute the call option's pay-off by subtracting the strike and taking the maximum with zero. The pay-off is then added to `runningSum` and the loop continues.

Once the loop is complete, we divide by the number of paths to get the expectation. Finally, we discount to get our estimate of the price which we return.

The `main` program takes in the inputs from the user, calls the Monte Carlo function, and displays the results. It asks for a final input to stop the routine from returning before the user has had a chance to read the results.

1.4 Critiquing the simple Monte Carlo routine

The routine we have written runs quickly and does what it was intended to do. It is a simple straightforward procedural program that performs as required. However, if we worked with this program we would swiftly run into annoyances. The essence of good coding is reusability. What does this mean? One simple definition is that code is reusable if someone has reused it. Thus reusability is as much a social concept as a technical one. What will make it easy for someone to reuse your code? Ultimately, the important attributes are clarity and elegance of design. If another coder decides that it would take as much effort to recode your routines as to understand them, then he will recode, and his inclination will be to recode in any case, as it is more fun than poring over someone else's implementation.

The second issue of elegance is equally important. If the code is clear but difficult to adapt then another coder will simply abandon it, rather than put lots of effort into forcing it to work in a way that is unnatural for how it was built.

The demands of reusability therefore mean we should strive for clarity and elegance. In addition, we should keep in mind when considering our original design the possibility that in future our code might need to be extended.

We return to our simple Monte Carlo program. Suppose we have a boss and each day he comes by and asks for our program to do something more. If we have designed it well then we will simply have to add features; if we have designed poorly then we will have to rewrite existing code.

So what might the evil boss demand?

"Do puts as well as calls!"

"I can't see how accurate the price is, put in the standard error."

"The convergence is too slow, put in anti-thetic sampling."

"I want the most accurate price possible by 9am tomorrow so set it running for 14 hours."

"It's crucial that the standard error is less 0.0001, so run it until that's achieved. We're in a hurry though so don't run it any longer than strictly necessary."

"I read about low-discrepancy numbers at the weekend. Just plug them in and see how good they are."

"Apparently, standard error is a poor measure of error for low-discrepancy simulations. Put in a convergence table instead."

"Hmm, I miss the standard error can we see that too."

"We need a digital call pricer now!"

"What about geometric average Asian calls?"

"How about arithmetic average Asian puts?"

"Take care of variable parameters for the volatility and interest rates."

"Use the geometric Asian option as a control variate for the arithmetic one."

"These low-discrepancy numbers apparently only work well if you Brownian bridge. Put that in as well."

"Put in a double digital geometric Asian option."

"What about jump risk? Put in a jump-diffusion model."

To adapt the routine as written would require a rewrite to do any of these. We have written the simplest routine we could think of, without considering design issues. This means that each change is not particularly natural and requires extra work.

For example, with this style of programming how would we would do the put option?

Option one: copy the function, change the name by adding put at the end, and rewrite the two lines where the pay-off is computed.

Option two: pass in an extra parameter, possibly as an `enum` and compute the pay-off via a `switch` statement in each loop of the Monte Carlo. The problem with the first option is that when we come to the next task, we have to adapt both the functions in the same way and do the same thing twice. If we then need more pay-offs this will rapidly become a maintenance nightmare.

The issues with the other option are more subtle. One problem is that a switch statement is an additional overhead so that the routine will now run a little slower. A deeper problem is that when we come to do a different routine which also uses a pay-off, we will have to copy the code from inside the first routine or rewrite it as necessary. This once again becomes a maintenance problem; every time we want to add a new sort of pay-off we would have to go through every place where pay-offs are used and add it on.

A C style approach to this problem would be to use a function pointer, we pass a pointer to a function as an argument to the Monte Carlo. The function pointed to is then called via the pointer in each loop to specify the price. Note that the call to the function would have to specify the strike as well as spot since the function could not know its value. Note also that if we wanted to do a double-digital option we would have problems as the double digital pays if and only if spot is between two levels, and we only have one argument, the strike, to play with.

The C++ approach to this problem is to use a class. The class would encapsulate the behaviour of the pay-off of a vanilla option. A pay-off object would then be passed into the function as an argument and in each loop a method expressing its value would be called to output the price for that pay-off. We look at the implementation of such a class in the next chapter.

1.5 Identifying the classes

In the previous section, we saw that the problem of wanting to add different sorts of vanilla options led naturally to the use of a class to encapsulate the notion of a pay-off. In this section, we look at identifying other natural classes which arise from the boss's demands.

Some of the demands were linked to differing forms that the boss wanted the information in. We could therefore abstract this notion by creating a statistics gatherer class.

We also had differing ways of terminating the Monte Carlo. We could terminate on time, on standard error or simply after a fixed number of paths. We could abstract this by writing a terminator class.

There were many different issues with the method of random number generation. The routine as it stands relies on the inbuilt generator which we do not know much about. We therefore want to be able to use other random number generators. We also want the flexibility of using low-discrepancy numbers which means another form of generation. (In addition, Box–Muller does not work well with low-discrepancy numbers so we will need flexibility in the inputs.) Another natural abstraction is therefore a random number generator class.

As long as our option is vanilla then specifying its parameters via pay-off and strike is fairly natural and easy; however, it would be neater to have one class that contains both pieces of information. More generally, when we pass to path-dependent exotic options, it becomes natural to have a class that expresses the option's properties. What would we expect such a class to do? Ultimately, an easy way to decide what the class should and should not know is to think of whether a piece of information would be contained in the term-sheet. Thus the class would know the pay-off of the option. It would know the expiry time. If it was an Asian it would know the averaging dates. It would also know whether the averaging was geometric or arithmetic. It would not know anything about interest rates, nor the value of spot nor the volatility of the spot rate as none these facts are contained in the term-sheet. The point here is that by choosing a real-world concept to encapsulate, it is easy to decide what to include or not to include. It is also easy for another user to understand how you have decided what to include or not to include.

What other concepts can we identify? The concept of a variable parameter could be made into a class. The process from which spot is drawn is another concept. The variable interest rates could be encapsulated via a class that expresses the notion of a discount curve.

1.6 What will the classes buy us?

Suppose that having identified all these classes, we implement them. What do we gain?

The first gain is that because these classes encapsulate natural financial concepts, we will need them when doing other pieces of coding. For example, if we have a class that does yield curves then we will use it time and time again, as to price any derivative using any reasonable method involves knowledge of the discount curve. Not only will we save time on the writing of code but we will also save time on the debugging. A class that has been tested thoroughly once has been tested forever and in addition, any little quirks that evade the testing regime will be found through repeated reuse. The more times and ways something has been reused the fewer the bugs that will be left. So using reusable classes leads to more reliable code as well as saving us coding time. Debugging often takes at least as much time as coding in any case, so saving time on debugging is a big benefit.

A second gain is that our code becomes clearer. We have written the code in terms of natural concepts, so another coder can identify the natural concepts and pick up our code much more easily.

A third gain is that the classes will allow us to separate interface from implementation. All the user needs to know about a pay-off class or discount curve class are what inputs yield what outputs? How the class works internally does not matter. This has multiple advantages. The first is that the class can be reused without the coder having to study its internal workings. The second advantage is that because the defining characteristic of the class is what it does but not how it does it, we can change how it does it at will. And crucially, we can do this without rewriting the rest of our program. One aspect of this is that we can first quickly write a suboptimal implementation and improve it later at no cost. This allows us to provide enough functionality to test the rest of the code before devoting a lot of time to the class. A third advantage of separating interface from implementation is that we can write multiple classes that implement the same interface and use them without rewriting all the interface routines. This is one of the biggest advantages of object-oriented design.

In the next chapter, we look at some of these concepts in the concrete case of a pay-off class.

1.7 Why object-oriented programming?

This is a book about implementing pricing models using object-oriented C++ programs. The reader may ask why is this worth learning? A short answer is that this is the skill you need if you want a job working as a quantitative analyst or quantitative developer. But this begs the question of why is this the required skill?

Object-oriented programming has become popular as computer projects have become larger and larger. A single project may now involve millions of lines of code. No single programmer will ever be able to hold all of that code in his mind at once. Object-oriented programming provides us with a way of coding that corresponds to natural mental maps. We know what each class of objects does, and more importantly we tightly define how they can interact with each other. This allows a clear map in the coder's mind of how the code fits together. And equally importantly, this allows easy communication of the code's structure to other programmers in the team.

When the coder needs to focus in on a particular part of the code, he need only look at the interior of the particular object involved and its interface with other objects. As long as the interface is not broken, and the new object lives up to the same responsibilities as the old one then there is no danger of unexpected ramifications (i.e. bugs) in distant parts of the code. Thus object-oriented programming leads to more robust code that is easier for teams to work on.

1.8 Key points

In this chapter, we have looked at how to implement a simple Monte Carlo routine on a procedural program. We then criticized it from the point of view of easy extensibility and reuse.

- Options can be priced by risk-neutral expectation.
- Monte Carlo uses the Law of Large Numbers to approximate this risk-neutral expectation.
- Reuse is as much a social issue as a technical one.
- Procedural programs can be hard to extend and reuse.
- Classes allow us to encapsulate concepts which makes reuse and extensibility a lot easier.
- Making classes closely model real-world concepts makes them easier to design and to explain.

- Classes allow us to separate the design of the interface from the coding of the implementation.

1.9 Exercises

Exercise 1.1 Modify the program given to price puts.

Exercise 1.2 Modify the program given to price double digitals.

Exercise 1.3 Change the program so that the user inputs a string which specifies the option pay-off.

Exercise 1.4 Identify as many classes as you can in the evil boss's list of demands.

2

Encapsulation

2.1 Implementing the pay-off class

In the last chapter, we looked at a simple Monte Carlo pricer and concluded that
the program would be improved if we used a class to encapsulate the notion of the
pay-off of a vanilla option. In this section, we look at how such a pay-off might be
implemented. In the files PayOff1.h and Payoff1.cpp, we develop such a class.
Before looking at the source file, PayOff1.cpp, we examine the more important
header file. The header file is much more important because it is all that other files
will see, and ideally it is all that the other files need to see.

Listing 2.1 (PayOff1.h)

```
#ifndef PAYOFF_H
#define PAYOFF_H


class PayOff
{
public:
    enum  OptionType {call, put};
    PayOff(double Strike_, OptionType TheOptionsType_);
    double operator()(double Spot) const;

private:
    double Strike;
    OptionType TheOptionsType;
};
#endif
```

We use an enum to distinguish between different sorts of pay-offs. If we ever
want more than put and call, we would add them to the list. We will discuss

13

more sophisticated approaches in Chapter 3 but this approach is enough to get us started.

The constructor

```
PayOff(double Strike_, OptionType TheOptionsType_)
```

takes in the strike of the option and the type of the option pay-off.

The main method of the class is

```
double PayOff::operator()(double spot) const
```

The purpose of this method is that given a value of spot, it returns the value of the pay-off.

Note the slightly odd syntax: we have overloaded `operator()`. To call this method for an object `thePayoff` with spot given by S we would write `thePayoff(S)`. Thus when use the object it appears more like a function than an object; such objects are often called function objects or functors. (Note that the C++ concept of functor is quite different from the pure mathematical one.)

We have defined `operator()` to be `const`. This means that the method does not modify the state of the object. This is as we would expect: computing the pay-off does not change the strike or the type of an option.

Suppose we did not specify that `operator()` was `const`; what would happen? The same functionality would be provided and the code would still compile. However, if a pay-off object was declared `const` at some point by a user then the compiler would complain if we tried to call `operator()`, and give us a complicated message to the effect that we cannot call non `const` methods of `const` objects. Thus if we want a method to be usable in `const` objects we must declare the method `const`.

An alternative approach, adopted by some programmers, is not to use `const` anywhere. The argument goes along the lines of "If we use `const` in some places, we must use it everywhere, and all it does is cause trouble and stop me doing what I want so why bother? Life will be much easier if we just do not use it."

Yet I use `const` as much as possible. The reason is that it is a safety enforcement mechanism. Thinking about `const` forces me to consider the context of an object and whether or not I wish it to change when doing certain things. If I am woolly in my thinking then the compiler will generally squeal when I attempt to compile, and thus errors are picked up at compile time instead of at run time.

A second benefit is that by giving the compiler the extra information that objects are `const`, we allow it to make extra optimizations. Code on a good compiler can therefore run faster when `const` is used. Now we are ready for `PayOff1.cpp`.

Listing 2.2 (PayOff1.cpp)

```
#include <PayOff1.h>
#include <MinMax.h>


PayOff::PayOff(double Strike_, OptionType TheOptionsType_)
:
    Strike(Strike_), TheOptionsType(TheOptionsType_)
{
}


double PayOff::operator ()(double spot) const
{
    switch (TheOptionsType)
    {
    case call :
        return max(spot-Strike,0.0);

    case put:
        return max(Strike-spot,0.0);

    default:
        throw("unknown option type found.");
    }
}
```

2.2 Privacy

We have declared the data in the pay-off class to be private. This means that the data cannot be accessed by code outside the class. The only code that can see, let alone change, their values are the constructor and the method operator(). What does this buy us? After all, for some reason the user might want to know the strike of an option and we have denied him the possibility of finding it out.

The reason is that as soon we let the user access the data directly, it is much harder for us to change how the class works. We can think of the class as a contract between coder and user. We, the coder, have provided the user with an interface: if he gives us spot we will give him the value of the pay-off. This is all we have contracted to provide. The user therefore expects and receives precisely that and no more.

For example, if the user could see the strike directly and accessed it, and if we changed the class so that the strike was no longer stored directly (which we shall do in the next chapter), then we would get compile errors from everywhere the strike was accessed. If the class had been used a lot in many files, in many different projects (which is after all the objective of reuse), then to find every place where strike had been accessed would be a daunting task. In fact, if this were the case we would probably consider finding them all a considerable waste of time, and we would probably give up reforming the internal workings.

Thus by making the data private, we can enforce the contract between coder and user in such a way that the contract does not say anything about the interior workings of the class. If for some reason, we wanted the user to be able to know the strike of the pay-off then we could add in an extra method `double GetStrike() const`. Whilst this would seem to undo all the benefits of using private data, this is not so since it still allows us the possibility of storing the data in a totally different way.

To conclude, by using private data we can rework the interior workings of a class as and when we need to without worrying about the impact on other code. That is, private data helps us to separate interface and implementation.

2.3 Using the pay-off class

Having designed a pay-off class, we want to use it in our Monte Carlo routines. We do this in the function `SimpleMonteCarlo2` which is declared in `SimpleMC.h` and defined in `SimpleMC.cpp`.

Listing 2.3 (`SimpleMC.h`)

```
#ifndef SIMPLEMC_H
#define SIMPLEMC_H
#include <PayOff1.h>

double SimpleMonteCarlo2(const PayOff& thePayOff,
                         double Expiry,
                         double Spot,
                         double Vol,
                         double r,
                         unsigned long NumberOfPaths);
#endif
```

Listing 2.4 (`SimpleMC.cpp`)

```
#include <SimpleMC.h>
#include <Random1.h>
#include <cmath>
```

```
// the basic math functions should be in
// namespace std but aren't in VCPP6
#if !defined(_MSC_VER)
using namespace std;
#endif

double SimpleMonteCarlo2(const PayOff& thePayOff,
                         double Expiry,
                         double Spot,
                         double Vol,
                         double r,
                         unsigned long NumberOfPaths)
{
   double variance = Vol*Vol*Expiry;
   double rootVariance = sqrt(variance);
   double itoCorrection = -0.5*variance;

   double movedSpot = Spot*exp(r*Expiry +itoCorrection);
   double thisSpot;
   double runningSum=0;

   for (unsigned long i=0; i < NumberOfPaths; i++)
   {
      double thisGaussian = GetOneGaussianByBoxMuller();
      thisSpot = movedSpot*exp( rootVariance*thisGaussian);
      double thisPayOff = thePayOff(thisSpot);
      runningSum += thisPayOff;
   }
   double mean = runningSum / NumberOfPaths;
   mean *= exp(-r*Expiry);
   return mean;
}
```

The main change from our original Monte Carlo routine is that we now input a
PayOff object instead of a strike. The strike is, of course, now hidden inside the
inputted object. We pass the object by reference and make it const as we have no
need to change it inside our routine. Note that the only line of our algorithm that is
new is

```
      double thisPayOff = thePayOff(thisSpot);
```

We illustrate how the routine might be called in `SimpleMCMain2.cpp`. Here we define both call and put objects and this shows that the routine can now be used without change to prices both types of options.

Listing 2.5 (`SimpleMCMain2.cpp`)

```
/*
requires

        PayOff1.cpp
        Random1.cpp,
        SimpleMC.cpp,
*/


#include<SimpleMC.h>
#include<iostream>
using namespace std;

int main()
{
        double Expiry;
        double Strike;
        double Spot;
        double Vol;
        double r;
        unsigned long NumberOfPaths;

        cout << "\nEnter expiry\n";
        cin >> Expiry;

        cout << "\nEnter strike\n";
        cin >> Strike;

        cout << "\nEnter spot\n";
        cin >> Spot;

        cout << "\nEnter vol\n";
        cin >> Vol;

        cout << "\nr\n";
        cin >> r;
```

```
cout << "\nNumber of paths\n";
cin >> NumberOfPaths;

PayOff callPayOff(Strike, PayOff::call);
PayOff putPayOff(Strike, PayOff::put);

double resultCall = SimpleMonteCarlo2(callPayOff,
                                      Expiry,
                                      Spot,
                                      Vol,
                                      r,
                                      NumberOfPaths);

double resultPut = SimpleMonteCarlo2(putPayOff,
                                     Expiry,
                                     Spot,
                                     Vol,
                                     r,
                                     NumberOfPaths);

cout <<"the prices are " << resultCall
                         << "  for the call and "
                         << resultPut
                         << " for the put\n";

double tmp;
cin >> tmp;

return 0;
}
```

2.4 Further extensibility defects

We have now set-up a pay-off class. One of our original objectives was to be able to easily extend the code without needing to rewrite other code when we add new pay-offs. Have we achieved this? Yes.

Suppose we now want to add in the digital call pay-off. What do we need to do? We add in an extra type, `digitalcall`, in the enum. We also change the `switch` statement in the `operator()` to include the extra case of a digital call and in that case return 1 if spot is above strike and 0 otherwise.

Everywhere the `PayOff` class has been used, in our case the Monte Carlo routine but possibly in many places, the digital call is now available just by passing in a `PayOff` object in an appropriate state.

Do this change for yourself and hit 'build'. One slightly non-obvious effect is that all the files which involve pay-offs are now recompiled. So although there is no obvious difference in those files, we see something slightly unexpected: as they include the file `PayOff1.h`, and it has changed, they must be recompiled too.

This is a warning sign that the code is not as independent as we would like. In an extreme case, we might not have access to source code for purchased libraries; any programming model that required us to recompile those libraries would be useless to us. In addition even if we could recompile, it might be a time-consuming process which we would prefer to avoid.

We would therefore like a way of programming that allows us not just to add functionality without modifying dependent files, but also to be able to do so without having to recompile existing files.

In fact, any solution that allowed us to do so would have to allow us to add an extra sort of pay-off without changing the pay-off class that the files using pay-offs sees. For if the part that they see changes in any way, the compiler will insist on recompiling them even if the changes do not appear to be material.

2.5 The open-closed principle

The previous section's discussion leads naturally to a programming idea known as the *open-closed principle*. The 'open' refers to the idea that code should always be open for extension. So in this particular case, we should always be able to add extra pay-offs.

The 'closed' means that the file is 'closed for modification'. This refers to the idea that we should be able to do the extension without modifying any existing code, and we should be able to do so without even changing anything in any of the existing files.

This may seem a little counterintuitive; how can one possibly make new forms of pay-offs without changing the pay-off class? To illustrate the idea before presenting the C++ solution, we consider how we might do this using C style techniques.

Suppose instead of making the class contain an `enum` that defines the option type, we instead use a function pointer. Thus we replace `OptionType` with

```
double (*FunctionPtr)(double,double).
```

The constructor for the pay-off would then take in the strike and a function pointer. It would store both and when `operator()` was called it would dereference the pointer and call the function pointed to with spot and strike as its arguments.

This code achieves a lot of our objectives. We can put the function pointed to in a new file, so the existing file for the pay-off class does not change each time we add a new form of pay-off. This means that neither the pay-off file nor the Monte Carlo file which includes the header file for pay-off need to be recompiled let alone changed.

However, there is still a fly in the ointment. What do we do when the boss comes by and demands a double-digital pay-off? The double digital requires two strikes (or barrier levels) and we only have one slot, the strike. Now this was also a problem with our original pay-off class; it too had only one slot for the strike.

One solution would be to use an array to specify and store parameters. The strike would be replaced by an array in both constructor and class members. The function pointer would take the array and spot as it arguments. This could be made to work. However, the code is now starting to lose the properties of clarity and elegance which were a large part of our original objectives.

So although the function pointer gets us a long way it does not get us far enough and we need a more sophisticated approach. The issue is that we need a way of specifying an object for the pay-off which is not of a predetermined type. This object will contain all the data it needs know and no more. So for a vanilla call or put the object would contain the strike, but for a double digital it would contain both of the two barrier levels.

Fortunately, C++ was designed with just this sort of problem in mind, and in the next chapter we shall study how to use inheritance and virtual functions to overcome these problems. We have used the open-closed principle as a way of motivating the introduction of these concepts, and we shall repeatedly return to it as a paradigm for evaluating programming approaches.

2.6 Key points

In this chapter, we looked at one way of using a class to encapsulate the notion of a pay-off. We then saw some of the advantages of using such a class. We also saw that the class had not achieved all of our objectives.

- Using a pay-off class allows us to add extra forms of pay-offs without modifying our Monte Carlo routine.
- By overloading `operator()` we can make an object look like a function.
- `const` enforces extra discipline by forcing the coder to be aware of which code is allowed to change things and which code cannot.
- `const` code can run faster.

- The open-closed principle says that code should be open for extension but closed for modification.
- Private data helps us to separate interface from implementation.

2.7 Exercises

Exercise 2.1 Modify the pay-off class so that it can handle digital options.

Exercise 2.2 Modify the pay-off class so that it can handle double-digital options.

Exercise 2.3 Test whether on your compiler using `const` speeds code up.

3

Inheritance and virtual functions

3.1 'is a'

We reconsider our `PayOff` class. At the end of the last chapter, we concluded that we would like to be able to use differing sorts of objects as pay-offs without changing any of the code that takes in pay-off objects. In other words, we would like to be able to say that a call option pay-off 'is a' pay-off, and express this idea in code.

The mechanism for expressing the 'is a' relationship in C++ is inheritance. There are plenty of examples we have already seen where such 'is a' relationships are natural. For example, a call option is a vanilla option. The compiler's built-in random number generator is a random number generator. Box–Muller is a method of turning uniform random variables into Gaussian random variables. An Asian option is a path-dependent exotic option. An arithmetic Asian call option is an Asian option, as is a geometric Asian put option. Returning the mean is a way of gathering statistics for a Monte Carlo simulation. Specifying the continuously compounding rate is a way to specify the discount curve. The BlackScholes model is a model of stock price evolution.

Thus 'is a' relationships are very natural in mathematical finance (as well as in coding and life in general.) We shall use inheritance to express such relationships. We shall refer to the class we inherit from as the *base class* and the class which does the inheriting will be called the *inherited class*. The base class will always express a more general concept than the inherited class.

Note that there is nothing to stop us inheriting several times. For example, our base might be `PathDependentExoticOption`. An inherited class might be `AsianOption`. We might then further inherit `AsianPutOption` and `AsianCallOption` from `AsianOption`.

The key point is that each inherited class refines the class above it in the hierarchy to something more specific.

3.2 Coding inheritance

To indicate that a class, `PayOffCall`, is inherited from a class `PayOff`, we use the key word `public`. Thus when declaring the inherited class, we write

```
class PayOffCall : public PayOff
```

What effect does this have? `PayOffCall` inherits all the data members and methods of `PayOff`. And most importantly, the compiler will accept a `PayOffCall` object wherever it expects a `PayOff` object. Thus we can write all our code to work off `PayOff` objects but then use inherited class objects to specify the precise properties. (There are some issues, however; see Section 3.4.)

We can also add data members and methods at will. The rules of `public` inheritance say that we can access `protected` data and methods of the base class inside the methods of the inherited class but we cannot access the `private` data. It is generally recommended to use `private` data for this reason; if we did otherwise then any changes in the design of the base class might require that any inherited classes be rewritten and at the very least all inherited classes would have to be checked. By using `private` data, we encapsulate what the base class does, and allow ourselves to reimplement it in future. It is, however, safe to use `protected` methods, as these, similarly to `public` methods, define part of the object's interface. As long as the implicit (or ideally explicit) contract which expresses the method's functionality does not change, it does not matter what interior changes are made.

3.3 Virtual functions

Returning to our example of the `PayOff` class, we redefine the class to work using inheritance. Our base class is still called `PayOff` but whereas previously it did a lot, it now does almost nothing. In fact, it does one thing; it specifies an interface.

We give the code in `PayOff2.h` and `PayOff2.cpp`.

Listing 3.1 (`PayOff2.h`)

```
#ifndef PAYOFF2_H
#define PAYOFF2_H


class PayOff
{
public:
    PayOff(){};
    virtual double operator()(double Spot) const=0;
    virtual ~PayOff(){}
```

```
private:
};


class PayOffCall : public PayOff
{
public:
    PayOffCall(double Strike_);
    virtual double operator()(double Spot) const;
    virtual ~PayOffCall(){}

private:
    double Strike;

};


class PayOffPut : public PayOff
{
public:
    PayOffPut(double Strike_);
    virtual double operator()(double Spot) const;
    virtual ~PayOffPut(){}

private:
    double Strike;
};
#endif
```

Listing 3.2 (PayOff2.cpp)

```
#include <PayOff2.h>
#include <minmax.h>

PayOffCall::PayOffCall(double Strike_) : Strike(Strike_)
{
}
double PayOffCall::operator () (double Spot) const
{
    return max(Spot-Strike,0.0);
}
```

```
double PayOffPut::operator () (double Spot) const
{
    return max(Strike-Spot,0.0);
}


PayOffPut::PayOffPut(double Strike_) : Strike(Strike_)
{
}
```

The main changes to the pay-off class are that we have added the keyword virtual to operator() and we placed an =0, at the end of operator(). We have added in a destructor which is also virtual. We have also abolished all the data from both the constructor and the class definition.

We also have two new classes PayOffCall and PayOffPut. Each of these have been inherited from the class PayOff. These classes will now do all the work. The call pay-off 'is a' pay-off, and we will use PayOffCall instead of the pay-off class where we need a call pay-off. Similarly for the put pay-off.

The crucial point is that the main method, operator() is now *virtual*. What is a virtual function? In technical terms, it is a function whose address is bound at runtime instead of at compile time. What does that mean? In the code, where a PayOff object has been specified, for example in our simple Monte Carlo routine, the code when running will encounter an object of a class that has been inherited from PayOff. It will then decide what function to call on the basis of what type that object is. So if the object is of type PayOffCall, it calls the operator() method of PayOffCall, and if it is of type PayOffPut, it uses the method from the PayOffPut class and so on.

It's worth understanding how the compiler implements this. Rather than saying, "hmm, what type is this object? Ah, it's a call so I'll use the call pay-off function," it adds extra data to each object of the class which specifies what function to use. In fact, essentially what it stores is a function pointer. So virtual functions are really a fancy way of using function pointers. Indeed, if you run a program involving virtual function pointers through a debugger, and examine through the watch window an object from a class with virtual functions, you will find an extra data member, the virtual function table. This virtual function table is a list of addresses to be called for the virtual functions associated with the class. So if when running a program, a virtual function is encountered, the table is looked up and execution jumps into the function pointed to. Note that this operation takes a small amount of time so efficiency fanatics dislike virtual functions as they cause a small decrease in execution speed. Note also that the amount of memory per object has also increased as the object now contains a lot of extra data.

If virtual functions are just function pointers, why bother with them? The first reason is that they are syntactically a lot simpler. The structure of our program is much clearer: if we can say this is a pay-off call object and this is a pay-off put object rather than having to say that we have a pay-off object but it contains a pointer to a function defining a call pay-off, we have a gain in clarity.

A second and important reason is that we get extra functionality. Depending on the complexity of the pay-off we may need extra data which cannot be expressed by a single number. With inheritance, we simply require the inherited class to contain all the data necessary. Thus for a double-digital pay-off we simply have two `doubles` expressing the upper and lower barriers. For a power option, we have a `double` for the strike and an `unsigned long` for the power. We could even have some complicated object stored as an extra argument. Indeed, if we wanted to do a complicated pay-off as a linear combination of call options, the extra data could be further call options whose pay-offs would be evaluated inside the `operator()` method and added together.

As well as being a virtual function, the `operator()` method has an =0 after it. This means that it is a *pure* virtual function. A pure virtual function has the property that it need not be defined in the base class and must be defined in an inherited class. Thus by putting =0 we are saying that the class is incomplete, and that certain aspects of its interface must be programmed in an inherited class.

In `SimpleMCMain3.cpp` we give an example of how to use the new pay-off class. Note that it uses `SimpleMC2.h` which only differs from `SimpleMC.h` in the replacement of `PayOff.h` with `PayOff2.h`. So we omit it.

Listing 3.3 (`SimpleMCMain3.cpp`)

```
/*
requires
        PayOff2.cpp
        Random1.cpp
        SimpleMC2.cpp
*/
#include<SimpleMC2.h>
#include<iostream>
using namespace std;

int main()
{
    double Expiry;
    double Strike;
    double Spot;
```

```
double Vol;
double r;
unsigned long NumberOfPaths;

cout << "\nEnter expiry\n";
cin >> Expiry;

cout << "\nEnter strike\n";
cin >> Strike;

cout << "\nEnter spot\n";
cin >> Spot;

cout << "\nEnter vol\n";
cin >> Vol;

cout << "\nr\n";
cin >> r;

cout << "\nNumber of paths\n";
cin >> NumberOfPaths;

PayOffCall callPayOff(Strike);
PayOffPut putPayOff(Strike);

double resultCall = SimpleMonteCarlo2(callPayOff,
                                     Expiry,
                                     Spot,
                                     Vol,
                                     r,
                                     NumberOfPaths);

double resultPut = SimpleMonteCarlo2(putPayOff,
                                     Expiry,
                                     Spot,
                                     Vol,
                                     r,
                                     NumberOfPaths);

cout <<"the prices are " << resultCall <<
                          " for the call and "
```

```
                    << resultPut <<
                    " for the put\n";

    double tmp;
    cin >> tmp;

    return 0;
}
```

3.4 Why we must pass the inherited object by reference

In our Monte Carlo routine, we have a parameter of type `const PayOff&` called `thePayOff`. This means that the parameter is being passed by reference rather than by value. The routine therefore works off the original object passed in. If we had not used the '&' it would copy the object: it would be passed by value not by reference. Suppose we change the parameter to type `PayOff` with or without the `const`, what happens? The code will not compile.

Why not? When the argument `Payoff` is encountered, the compiler refuses to create an object of type `PayOff` because it has a pure virtual method. The method `operator()` has not been defined, and the rules of C++ say that you cannot create an object of a type with a pure virtual method.

How could we get round this? Suppose we decide to make `operator()` an ordinary virtual function by defining it. For example, we could just make it return 0 and still override in the inherited clases as before. The code with the '&' would compile and run as before, giving the same results. The code without the '&' would now compile and run. However, the price of every option would be zero, which is certainly not what we want.

This happens because when the compiler encounters the argument of type `PayOff`, the input parameter is copied into a variable of type `PayOff`. This occurs because the compiler will call the copy constructor of `PayOff`. Like all copy constructors, it takes in an object of type `const PayOff&`. As the compiler happily accepts references to inherited objects instead of references to base-class objects this is legal and the compiler does not complain. However, the copy constructor of `PayOff` is not interested in all the extra data and information contained in the inherited object: it just looks down the data coming from the base class and copies it into the new object. As the new object is truly a base class object rather than an inherited class object pretending to be a base class object, its virtual functions will be those of the base class.

This all means that the new variable has the size and data of a `PayOff` object regardless of the type of the inputted object. The compiler therefore truncates all the additional data members which have been added, and the virtual function table is that of the base class object not the inherited class. In fact, disastrous things would occur if the new object inherited the virtual function table of the inherited object, as the virtual methods would try to access non-existent data members with possibly dangerous consequences.

Making the base class method concrete instead of pure virtual was therefore a mistake, and, in fact, the compiler's rejection of the argument without the '&' was saving us from a dangerous error.

It's worth thinking a little about what actually happens when the object is passed by reference. All that happens is that the function is passed the address of the object in memory, no copying occurs and the object's state is precisely as it was outside. However, if the object's state does change inside the routine it will also change outside which may not be what we want. We therefore include the `const` to indicate that the routine cannot do anything which may change the state of the object. The function can 'look but not touch.'

3.5 Not knowing the type and virtual destruction

Generally, one cannot pass an object of one type where another type is expected; a big feature of C++ is that it enforces type security. This means that the compiler picks up many errors at compile time instead of at run time. We cannot even run our program until we have made sure that all objects are of the types expected. This reduces the number of bugs that need to be found while the program is running by enforcing some discipline.

However, the rules of inheritance say that we can always pass a reference to an object of an inherited class instead of a reference to a base class object. This means that at times we do not know the type of object. For example, inside the Monte Carlo routine, `SimpleMC2`, we appear to be using a base class object not an inherited one.

There are two ways to forget the type of the object. The first, which we have already used, is to refer to it via a reference to the base class type instead of the inherited class type. The second, related, method is via a pointer.

If we have a pointer to an inherited object we can always cast it into a base class object without any trouble. We give an example of this in `SimpleMC-Main4.cpp`.

Listing 3.4 (`SimpleMCMain4.cpp`)

```
/*
requires
        PayOff2.cpp
```

```
        Random1.cpp
        SimpleMC2.cpp
*/
#include<SimpleMC2.h>
#include<iostream>
using namespace std;

int main()
{
    double Expiry;
    double Strike;
    double Spot;
    double Vol;
    double r;
    unsigned long NumberOfPaths;

    cout << "\nEnter expiry\n";
    cin >> Expiry;

    cout << "\nEnter strike\n";
    cin >> Strike;

    cout << "\nEnter spot\n";
    cin >> Spot;

    cout << "\nEnter vol\n";
    cin >> Vol;

    cout << "\nr\n";
    cin >> r;

    cout << "\nNumber of paths\n";
    cin >> NumberOfPaths;

    unsigned long optionType;

    cout << "\nenter 0 for call, otherwise put ";
    cin >> optionType;

    PayOff* thePayOffPtr;
```

```
    if (optionType== 0)
        thePayOffPtr = new PayOffCall(Strike);
    else
        thePayOffPtr = new PayOffPut(Strike);

    double result = SimpleMonteCarlo2(*thePayOffPtr,
                                      Expiry,
                                      Spot,
                                      Vol,
                                      r,
                                      NumberOfPaths);

    cout <<"\nthe price is " << result << "\n";
    double tmp;
    cin >> tmp;

    delete thePayOffPtr;

    return 0;
}
```

The user of our unsophisticated interface enters a zero for a call option and a non-zero number for a put option. We use an `if` statement to create the pay-off object. Note the important point here is that we use `new`. Whilst we would like to simply write

```
if (optionType== 0)
  PayOffCall thePayOff(Strike);
else
  PayOffPut thePayOff(Strike);
```

this will not give us what we want. The reason is that whilst the object thePayOff will happily be created, it will be an *automatic variable*, which is one that automatically disappears whenever the *scope* is left. In this case, the scope is the dependent clause of the `if` statement, so as soon we leave the `if` statement the object no longer exists and any attempts to reference thePayOff will result in compiler errors.

On the other hand, when we use `new` we are instructing the compiler that we wish to take some memory whilst the code is running, and that memory should not be released until we explicitly say so. The object created will therefore continue to exist outside the `if` statement as desired. The operator `new` finds enough memory to put the object into and calls the constructor, placing the object at the memory that has been allocated. It returns a pointer to the object created.

Thus the code `new PayOffCall(Strike)` creates the `PayOffCall` object and it returns a pointer to the object so we can access it. The returned pointer will be of type `PayOffCall*`. Fortunately, we can always cast a pointer from an inherited class pointer into a base class pointer. We therefore assign `payOffPtr` to the result of `new` and we have the desired result. Note that we declare `payOffPtr` outside the `if` to ensure that it persists after the `if` statement is complete.

When we reach the call to the Monte Carlo routine, we can now pass in either a call or a put, depending on what the pointer points to. We dereference the pointer in order to pass in the object rather than the pointer by putting `*payOffPtr`.

The final thing we must do is get rid of the object. By using `new`, we told the compiler that it must not destroy the object nor deallocate the memory until we say so. If we never tell it to do so then the memory will never be freed up, and our memory will slowly leak away until program crashes. The way we instruct the compiler to destroy the object and deallocate the memory is to use the operator `delete`. When we call `delete`, it first calls the destructor of the object. At this point we must be careful: we have a pointer to a base object, so which destructor will it call? If the destructor is not `virtual` then it will call the base class destructor. If the object is of an inherited class this may cause problems as the object will not be destroyed properly. For example, if the inherited class object had an array as a data member then the memory allocated for that array will never be deallocated. In our case, the base class is abstract and there cannot be any objects of its type. This means that calling the base class destructor must be an error. For this reason, we declare the destructor `virtual`. The compiler then uses the virtual function table to correctly decide which destructor to call. In fact, many compilers issue a warning if you declare a method pure virtual and do not declare the destructor virtual.

In fact, when calling and executing the destructor of an inherited class, the compiler always calls the base class destructor; this ensures that all the data members of the object which arise from the base class are correctly destroyed.

In this section, we looked at one case where we did not know the type of an object and this caused us a little trouble, but it was also very useful. By treating the inherited class object as a base class object, we were able to make the same code work regardless of the precise type of the object. The important fact was that whatever the type of the object, it had to implement all the methods defined in the base class and this was enough to ensure that the code ran smoothly. One way of summarizing this situation is to say that the inherited class *implements the interface* defined by the base class.

In the next chapter, we will look at some additional problems caused by not knowing types, such as the problem of copying an object of unknown type, and examine their solutions.

3.6 Adding extra pay-offs without changing files

One of our objectives when rewriting the `PayOff` class was to create a class that could be extended without changing any of the existing code. In addition, we wished to be able to add extra pay-offs without having to recompile either the file containing the `PayOff` class or any files which included the file defining the `PayOff` class. In this section, we see how to do this with our class definition.

Suppose the new pay-off we wish to add is the double digital pay-off. This pay-off pays 1 if spot is between two values and 0 otherwise. We define the new pay-off class in a new file, `DoubleDigital.h` with the associated code in `DoubleDigital.cpp`.

Listing 3.5 (`DoubleDigital.h`)

```
#ifndef DOUBLEDIGITAL_H
#define DOUBLEDIGITAL_H
#include <Payoff2.h>

class PayOffDoubleDigital : public PayOff
{
public:
    PayOffDoubleDigital(double LowerLevel_,
                        double UpperLevel_);

    virtual double operator()(double Spot) const;
    virtual ~PayOffDoubleDigital(){}

private:
    double LowerLevel;
    double UpperLevel;
};
#endif
```

and the source code is

Listing 3.6 (`DoubleDigital.cpp`)

```
#include <DoubleDigital.h>
```

```
PayOffDoubleDigital::PayOffDoubleDigital(double LowerLevel_,
                                          double UpperLevel_)
                  :    LowerLevel(LowerLevel_),
                       UpperLevel(UpperLevel_)
{
}

double PayOffDoubleDigital::operator()(double Spot) const
{
    if (Spot <= LowerLevel)
        return 0;
    if (Spot >= UpperLevel)
        return 0;

    return 1;
}
```

The crucial point is that whilst we must include the file `DoubleDigital.h` in our interface we do not include it in either the Monte Carlo file, `SimpleMC2`, or the pay-off file, `PayOff2`. The changes to the interface file are minimal and we have

Listing 3.7 (`SimpleMCMain5.cpp`)

```
/*
requires DoubleDigital.cpp
         PayOff2.cpp
         Random1.cpp
         SimpleMC2.cpp
*/
#include<SimpleMC2.h>
#include<DoubleDigital.h>
#include<iostream>
using namespace std;

int main()
{
    double Expiry;
    double Low,Up;
    double Spot;
```

```
    double Vol;
    double r;
    unsigned long NumberOfPaths;

    cout << "\nEnter expiry\n";
    cin >> Expiry;

    cout << "\nEnter low barrier\n";
    cin >> Low;

    cout << "\nEnter up barrier\n";
    cin >> Up;

    cout << "\nEnter spot\n";
    cin >> Spot;

    cout << "\nEnter vol\n";
    cin >> Vol;

    cout << "\nr\n";
    cin >> r;

    cout << "\nNumber of paths\n";
    cin >> NumberOfPaths;

    PayOffDoubleDigital thePayOff(Low,Up);

    double result = SimpleMonteCarlo2(thePayOff,
                                      Expiry,
                                      Spot,
                                      Vol,
                                      r,
                                      NumberOfPaths);

    cout <<"\nthe price is " << result << "\n";
    double tmp;
    cin >> tmp;

    return 0;
}
```

Thus adding the new pay-off only required recompilation of one file rather than many and this achieves one objective: the PayOff class is open for extension but closed for modification. In particular, we could have added the new pay-off even if we did not have access to the source code of the PayOff class.

What we would really like is to be able to add new pay-offs without changing any files. That is we would like to be able to add new pay-offs without changing the interfacing file. In the interface, the user would simply type the name of the pay-off and this would be translated into the relevant pay-off. Rather surprisingly it is possible to do this. The solution is to use a design known as the *Factory pattern*. We shall discuss how to do this in Chapter 10.

3.7 Key points

In this chapter, we looked at how inheritance could be used to implement a PayOff class that is closed for modification but open for extension.

- Inheritance expresses 'is a' relationships.
- A virtual function is bound at run time instead of at compile time.
- We cannot have objects from classes with pure virtual functions.
- We have to pass inherited class objects by reference if we do not wish to change the virtual functions.
- Virtual functions are implemented via a table of function pointers.
- If a class has a pure virtual functions then it should have a virtual destructor.

3.8 Exercises

Exercise 3.1 Write an inherited class that does power options, and use it to price some.

Exercise 3.2 Implement an interface in which the user inputs a string and this is turned into a pay-off class.

Exercise 3.3 In the evil boss's list of demands in Chapter 1, try to identify as many inheritance relationships as possible.

4

Bridging with a virtual constructor

4.1 The problem

We have written a simple Monte Carlo program which uses a polymorphic class PayOff to determine the pay-off of the vanilla option to be priced. If we think about the real-world objects we wish to model, a very natural object is the option itself. At the moment, we have two pieces of data, the expiry and the pay-off, it would be much more natural to have a single piece of data, the option, which encompassed both.

How would we do this? One simple solution is simply to copy all the pay-off code we have written, add an extra data member, Expiry, to the base class and an extra method, GetExpiry, to the base class and be done. However, this rather spoils the paradigm of code reuse – we would like to reuse the same code rather than cut and paste it. In addition, if we do this for each new class of options, then it will be very time consuming to add new types of pay-offs as we will have to write a new inherited class for each option type.

A preferable solution would be to define a Vanilla Option class which has as data members a PayOff object and a double to represent expiry. However, if we try this we immediately hit a snag; the class PayOff is abstract. The compiler will squeal if we attempt to have a data member for our class of type PayOff, as you cannot instantiate an object from an abstract class. The issues we encounter here are very similar to those of Section 3.4. As there, making the PayOff class non-abstract by defining operator() in the base class causes more trouble than it solves. Any attempt to make the data member equal to an inherited class member will simply result in the object being copied and truncated into a base class object which is not what we want.

What we want is for the Vanilla Option object to be able to contain an object from an unknown class. Note that as the size of the unknown class object will not be constant, we will not be able to put it directly in the Vanilla Option object as this would lead to Vanilla Option objects being of variable size. This is something that is not allowed in C++. However, it is possible to refer to extra data outside the object using pointers or references. This is after all what an array, or a list, or a string, does.

One solution is therefore to store a reference to a pay-off object instead of a pay-off object.

4.2 A first solution

In this section, we implement a Vanilla Option class using a reference to a pay-off object and then discuss what's wrong with it.

Listing 4.1 (Vanilla1.h)

```
#ifndef VANILLA_1_H
#define VANILLA_1_H

#include <PayOff2.h>

class VanillaOption
{
public:
    VanillaOption(PayOff& ThePayOff_, double Expiry_);
    double GetExpiry() const;
    double OptionPayOff(double Spot) const;

private:
    double Expiry;
    PayOff& ThePayOff;
};
#endif
```

The source file is

Listing 4.2 (Vanilla1.cpp)

```
#include <Vanilla1.h>

VanillaOption::VanillaOption(PayOff&ThePayOff_,
                  double Expiry_)
            : ThePayOff(ThePayOff_), Expiry(Expiry_)
{
}
```

```
double VanillaOption::GetExpiry() const
{
    return Expiry;
}


double VanillaOption::OptionPayOff(double Spot) const
{
    return ThePayOff(Spot);
}
```

This is all very straightforward. The only subtlety is that the class member data is of type PayOff& instead of PayOff. However, we can use a reference to a pay-off object in precisely the same way as a pay-off object so this causes no extra coding. Our class provides two methods, one giving the expiry of the option and the other stating the pay-off at expiry given spot.

We can now rewrite the Monte Carlo routine to use the VanillaOption class. We thus have

Listing 4.3 (SimpleMC3.h)

```
#ifndef SIMPLEMC3_H
#define SIMPLEMC3_H
#include <Vanilla1.h>


double SimpleMonteCarlo3(const VanillaOption& TheOption,
                        double Spot,
                        double Vol,
                        double r,
                        unsigned long NumberOfPaths);
#endif
```

and

Listing 4.4 (SimpleMC3.cpp)

```
#include<SimpleMC3.h>
#include <Random1.h>
#include <cmath>
// the basic math functions should be in namespace
// std but aren't in VCPP6
#if !defined(_MSC_VER)
using namespace std;
#endif
```

```cpp
double SimpleMonteCarlo3(const VanillaOption& TheOption,
                         double Spot,
                         double Vol,
                         double r,
                         unsigned long NumberOfPaths)
{
    double Expiry = TheOption.GetExpiry();

    double variance = Vol*Vol*Expiry;
    double rootVariance = sqrt(variance);
    double itoCorrection = -0.5*variance;

    double movedSpot = Spot*exp(r*Expiry +itoCorrection);
    double thisSpot;
    double runningSum=0;

    for (unsigned long i=0; i < NumberOfPaths; i++)
    {
        double thisGaussian = GetOneGaussianByBoxMuller();
        thisSpot = movedSpot*exp( rootVariance*thisGaussian);
        double thisPayOff = TheOption.OptionPayOff(thisSpot);
        runningSum += thisPayOff;
    }

    double mean = runningSum / NumberOfPaths;
    mean *= exp(-r*Expiry);
    return mean;
}
/*
```

Our new routine is very similar to the old one, the difference being that we pass in an option object instead of pay-off and an expiry separately. We obtain the expiry from the option via `GetExpiry()`.

Our main routine is then

Listing 4.5 (`VanillaMain1.cpp`)

```
/*
requires DoubleDigital.cpp
         PayOff2.cpp
         Random1.cpp
         SimpleMC3.cpp
```

```
        Vanilla1.cpp
*/

#include<SimpleMC3.h>
#include<DoubleDigital.h>
#include<iostream>
using namespace std;
#include<Vanilla1.h>

int main()
{
    double Expiry;
    double Low,Up;
    double Spot;
    double Vol;
    double r;
    unsigned long NumberOfPaths;

    cout << "\nEnter expiry\n";
    cin >> Expiry;

    cout << "\nEnter low barrier\n";
    cin >> Low;

    cout << "\nEnter up barrier\n";
    cin >> Up;

    cout << "\nEnter spot\n";
    cin >> Spot;

    cout << "\nEnter vol\n";
    cin >> Vol;

    cout << "\nr\n";
    cin >> r;

    cout << "\nNumber of paths\n";
    cin >> NumberOfPaths;

    PayOffDoubleDigital thePayOff(Low,Up);
```

```
    VanillaOption theOption(thePayOff, Expiry);

    double result = SimpleMonteCarlo3(theOption,
                                      Spot,
                                      Vol,
                                      r,
                                      NumberOfPaths);

    cout <<"\nthe price is " << result << "\n";
    double tmp;
    cin >> tmp;

return 0;
}
```

The main change is that now we first pass a PayOff object and the expiry time into a VanillaOption object and that is then passed into the Monte Carlo.

This program will compile and run but I do not like it. Why? The VanillaOption class stores a reference to a PayOff object which was defined outside the class. This means that if we change that object then the pay-off of the vanilla option will change. The vanilla option will not exist as independent object in its own right but will instead always be dependent on the PayOff object constructed outside the class. This is a recipe for trouble. The user of the VanillaOption will not expect changes to the PayOff object to have such an effect. In addition, if the PayOff object had been created using new as we did in the last chapter then it might be deleted before the option ceased to exist which would result in the vanilla option calling methods of a non-existent object which is bound to cause crashes.

Similarly, if the option was created using new then it would likely exist long after the PayOff object had ceased to exist, and we would get similarly dangerous behaviour.

4.3 Virtual construction

What do we really want to do? We want the vanilla option to store its own copy of the pay-off. However, we do not want the vanilla option to know the type of the pay-off object nor anything about any of its inherited classes for all the reasons we discussed in the last chapter. Our solution there was to use virtual functions: how can we use them here? Well the object knows its own type so it can certainly make a copy of itself. Thus we define a virtual method of the base class which causes the object to create a copy of itself and return a pointer to the copy.

Such a method is called a *virtual copy constructor*. The method is generally given the name `clone`. Thus if we want the `PayOff` class to be virtually copyable, we add a pure virtual method to the base class by

```
virtual PayOff* clone() const=0;
```

and define it in each inherited class. For example,

```
PayOff* PayOffCall::clone() const
{
    return new PayOffCall(*this);
}
```

Note that the statement `new PayOffCall(*this);` says create a copy of the object for which the clone method has been called, as the `this` pointer always points to the object being called. The call to `clone` is then a call to the copy constructor of `PayOffCall` which returns a copy of the original `PayOffCall`, and because the operator `new` has been used, we can be sure that the object will continue to exist.

Note that we have made the return type of clone `PayOff*` which is a pointer to the base class object. Strictly speaking according to the standard, we should be able to change the return type to `PayOffCall*`, as it is permissible according to the standard to change the return type of a virtual function from a base class pointer to a pointer to the inherited class. As inherited class pointers can always be converted into base class pointers, this does not cause problems. However many compilers will not compile this syntax, as this is an exception to the general rule that you cannot override the return type of a virtual function.

We give the revised `PayOff` class in `PayOff3.h`. We now have

Listing 4.6 (`PayOff3.h`)

```
#ifndef PAYOFF3_H
#define PAYOFF3_H

class PayOff
{
public:
    PayOff(){};

    virtual double operator()(double Spot) const=0;
    virtual ~PayOff(){}
    virtual PayOff* clone() const=0;
```

```
private:
};

class PayOffCall : public PayOff
{
public:
    PayOffCall(double Strike_);

    virtual double operator()(double Spot) const;
    virtual ~PayOffCall(){}
    virtual PayOff* clone() const;

private:
    double Strike;
};

class PayOffPut : public PayOff
{
public:
    PayOffPut(double Strike_);

    virtual double operator()(double Spot) const;
    virtual ~PayOffPut(){}
    virtual PayOff* clone() const;

private:
    double Strike;
};
#endif
```

and PayOff3.cpp

Listing 4.7 (PayOff3.cpp)

```
#include <PayOff3.h>
#include <minmax.h>

PayOffCall::PayOffCall(double Strike_) : Strike(Strike_)
{
}
```

```
double PayOffCall::operator () (double Spot) const
{
    return max(Spot-Strike,0.0);
}


PayOff* PayOffCall::clone() const
{
    return new PayOffCall(*this);
}


double PayOffPut::operator () (double Spot) const
{
    return max(Strike-Spot,0.0);
}


PayOffPut::PayOffPut(double Strike_) : Strike(Strike_)
{
}


PayOff* PayOffPut::clone() const
{
    return new PayOffPut(*this);
}
```

We similarly have to change the contents of DoubleDigital to reflect the extra method in the base class. We do this in DoubleDigital2.h and in DoubleDigital2.cpp, which we do not reproduce here.

Using PayOff3.h, we can now make copies of PayOff objects of unknown type and reimplement the VanillaOption class accordingly.

Listing 4.8 (Vanilla2.h)

```
#ifndef VANILLA_2_H
#define VANILLA_2_H


#include <PayOff3.h>


class VanillaOption
{
public:
    VanillaOption(const PayOff& ThePayOff_, double Expiry_);
```

```
    VanillaOption(const VanillaOption& original);
    VanillaOption& operator=(const VanillaOption& original);
    ~VanillaOption();

    double GetExpiry() const;
    double OptionPayOff(double Spot) const;

private:
    double Expiry;
    PayOff* ThePayOffPtr;
};
#endif
```

and the source file is

Listing 4.9 (Vanilla2.cpp)

```
#include <Vanilla2.h>

VanillaOption::VanillaOption(const PayOff&
                            ThePayOff_, double Expiry_)
                    :  Expiry(Expiry_)
{
    ThePayOffPtr = ThePayOff_.clone();
}

double VanillaOption::GetExpiry() const
{
    return Expiry;
}

double VanillaOption::OptionPayOff(double Spot) const
{
    return (*ThePayOffPtr)(Spot);
}

VanillaOption::VanillaOption(const VanillaOption& original)
{
    Expiry = original.Expiry;
    ThePayOffPtr = original.ThePayOffPtr->clone();
}
```

```
VanillaOption& VanillaOption::
                  operator=(const VanillaOption& original)
{
    if (this != &original)
    {
        Expiry = original.Expiry;
        delete ThePayOffPtr;
        ThePayOffPtr = original.ThePayOffPtr->clone();
    }
    return *this;
}


VanillaOption::~VanillaOption()
{
    delete ThePayOffPtr;
}
```

We modify `SimpleMC3` to get `SimpleMC4` by changing the name of the include file to be `Vanilla2.h` and doing nothing else.

Listing 4.10 (SimpleMC4.h)

```
#ifndef SIMPLEMC4_H
#define SIMPLEMC4_H

#include <Vanilla2.h>

double SimpleMonteCarlo3(const VanillaOption& TheOption,
                         double Spot,
                         double Vol,
                         double r,
                         unsigned long NumberOfPaths);

#endif
```

Listing 4.11 (SimpleMC4.cpp)

```
#include <SimpleMC4.h>
#include <Random1.h>
#include <cmath>

// the basic math functions should be in namespace std
// but aren't in VCPP6
```

```
#if !defined(_MSC_VER)
using namespace std;
#endif
double SimpleMonteCarlo3(const VanillaOption& TheOption,
                         double Spot,
                         double Vol,
                         double r,
                         unsigned long NumberOfPaths)
{

    double Expiry = TheOption.GetExpiry();

    double variance = Vol*Vol*Expiry;
    double rootVariance = sqrt(variance);
    double itoCorrection = -0.5*variance;

    double movedSpot = Spot*exp(r*Expiry +itoCorrection);
    double thisSpot;
    double runningSum=0;

    for (unsigned long i=0; i < NumberOfPaths; i++)
    {
        double thisGaussian = GetOneGaussianByBoxMuller();
        thisSpot = movedSpot*exp( rootVariance*thisGaussian);
        double thisPayOff = TheOption.OptionPayOff(thisSpot);
        runningSum += thisPayOff;
    }

    double mean = runningSum / NumberOfPaths;
    mean *= exp(-r*Expiry);
    return mean;
}
/*
```

Our main program is now VanillaMain2.cpp.

Listing 4.12 (VanillaMain2.cpp)

```
/*
requires PayOff3.cpp,
         Random1.cpp,
         SimpleMC4.cpp
         Vanilla2.cpp
*/
```

```cpp
#include<SimpleMC4.h>
#include<iostream>
using namespace std;
#include<Vanilla2.h>

int main()
{
    double Expiry;
    double Strike;
    double Spot;
    double Vol;
    double r;
    unsigned long NumberOfPaths;

    cout << "\nEnter expiry\n";
    cin >> Expiry;

    cout << "\nStrike\n";
    cin >> Strike;

    cout << "\nEnter spot\n";
    cin >> Spot;

    cout << "\nEnter vol\n";
    cin >> Vol;

    cout << "\nr\n";
    cin >> r;

    cout << "\nNumber of paths\n";
    cin >> NumberOfPaths;

    PayOffCall thePayOff(Strike);

    VanillaOption theOption(thePayOff, Expiry);

    double result = SimpleMonteCarlo3(theOption,
                            Spot,
                            Vol,
                            r,
                            NumberOfPaths);
```

```
    cout <<"\nthe call price is " << result << "\n";

    VanillaOption secondOption(theOption);

    result = SimpleMonteCarlo3(secondOption,
                               Spot,
                               Vol,
                               r,
                               NumberOfPaths);

    cout <<"\nthe call price is " << result << "\n";

    PayOffPut otherPayOff(Strike);
    VanillaOption thirdOption(otherPayOff, Expiry);
    theOption = thirdOption;

    result = SimpleMonteCarlo3(theOption,
                               Spot,
                               Vol,
                               r,
                               NumberOfPaths);

    cout <<"\nthe put price is " << result << "\n";

    double tmp;
    cin >> tmp;

return 0;
}
/*
```

The actual syntax in our `main` program has not changed much.

4.4 The rule of three

One difference between the `VanillaOption` class and all the classes we have previously defined is that we have included an assignment operator (i.e. `operator=`) and a copy constructor. The reason is that whilst the compiler by default generates these methods, the compiler's definitions are inadequate.

If we do not declare the copy constructor then the compiler will perform a *shallow copy* as opposed to a *deep copy*. A shallow copy means that the data members have simply been copied, whilst this is adequate when no memory allocation is required it will lead to disaster when it does. If we have shallow copied a Vanilla-Option then both copies of the object have the same member ThePayOff-Ptr, so any modifications to the object we have pointed-to will have the same effect in each copy. Whilst for our particular object this is not really an issue as we do not want to change it, in a more complicated class this would certainly lead to trouble.

More seriously, when the one of the two VanillaOption objects goes out of scope then its destructor will be called and the pointed-to PayOff object will be deleted. The object remaining will now be in trouble as attempts to use its pointer will lead to methods of a non-existent object being called. On top of which when the remaining object goes out of scope, its destructor will be called and there will be a second attempt to delete the pointed-to object. As the object has already been deleted, this will generally result in a crash.

Thus if we write a class involving a destructor, we will generally need a copy constructor too. In fact, we have very similar issues with the assignment operator so it is necessary to write it as well. This leads to the "rule of three," which states that if any one of destructor, assignment operator and copy constructor is needed for a class then all three are. Whilst it is possible to construct examples for which this is not true, it is a very good rule of thumb. If you feel that you will never want to copy or assign objects from a class and do not bother to write the appropriate methods then you are storing up trouble; when someone does use the copy constructor the compiler will not complain but the code will crash. One solution is to declare the copy constructor but make it private. The attempt to use it will then result in a compile-time error instead of a run-time error which would be more easily spotted and fixed. (Similarly for the assignment operator.)

We analyze the copy constructor from VanillaOption. It copies the value of Expiry from the original into the copy, naturally enough. For the pointer, we call the clone method generating a copy of the PayOff object which we store in the copy's pointer. Note that the data members of the original and its copy will not be equal, since the pointers will have different values. However, this is precisely what we want, as we do not want them to be using the original PayOff object.

Note, however, that if we decided to define a comparison operator (i.e. operator==) for the class, we would have to be careful to compare the objects pointed to rather than the pointers, or we would always get false.

Having analyzed the copy constructor, let's look at the assignment operator. The first thing to note is that its return type is VanillaOption&. The effect of this is that it is legitimate to code

```
a=b=c;
```

which is equivalent to

```
b=c;
a=b;
```

but saves a little typing. The second, odder, point is that it it also legitimate to code

```
(a=b) = c;
```

which is equivalent to

```
a=b;
a=c;
```

which is not particularly useful. If we made the return type `const` then the second of these examples would not compile which might be regarded as an advantage; however, for built-in types the return type has no `const` and it is generally better to be consistent. Note that if we had made the return type `void`, the first example would not compile. If we had made it `VanillaOption` rather than `Vanilla-Option&` then the two examples would both compile but give odd behaviour. The first would work but might result in extra copying of objects which would waste time. The second would end with a being equal to b, as the returned object from a=b would have nothing to do with a any more and would happily be assigned the value of c and then disappear.

The second point to note about the assignment operator is that the first thing we do is check against self-assignment. If we accidentally coded a=a, we would not want the code to crash. Whilst it might seem an unlikely occurence when objects are being accessed through pointers it is not immediately obvious whether two objects are actually the same object. By checking if the address of the assigning object is equal to the `this` pointer, we check against this eventuality and do nothing if it occurs as nothing is needed to be done to ensure that a is equal to a.

The rest of the assignment operator is really just a combination of destructor and copy constructor. The destructor part is needed to get rid of the `PayOff` which the object owned before, as otherwise it would never be deleted. The copy constructor part clones the pointed-to object so the new version of the assigned part has its own copy of the object as desired.

4.5 The bridge

We have now achieved what we desired: the vanilla option class can be happily copied and moved around just like any other class and it uses the code already written for the `PayOff` class so we have no unnecessary duplication. However, there is still one aspect of our implementation which is slightly annoying – we had to write special code to handle assignment, construction and destruction. Every time we want to write a class with these properties, we will have to do the same thing again and again. This is rather contrary to the paradigm of reuse. What we would really like is a pay-off class that has the nice polymorphic features that our class does, whilst taking care of its own memory management. There are a couple of approaches to solving this problem. One is to use a wrapper class that has

been templatized: this is really generic programming rather than object-oriented
programming, and we explore this approach in Section 5.3. The other is known
as *the bridge pattern*. Suppose we take the vanilla option class and get rid of the
member Expiry and the method GetExpiry. We then are left with a class that
does nothing except store a pointer to an option pay-off and takes care of memory
handling. This is precisely what we want.

   We present such a class:

Listing 4.13 (PayOffBridge.h)

```
#ifndef PAYOFFBRIDGE_H
#define PAYOFFBRIDGE_H

#include<PayOff3.h>

class PayOffBridge
{
public:
    PayOffBridge(const PayOffBridge& original);
    PayOffBridge(const PayOff& innerPayOff);

    inline double operator()(double Spot) const;
    ~PayOffBridge();
    PayOffBridge& operator=(const PayOffBridge& original);

private:
    PayOff* ThePayOffPtr;
};

inline double PayOffBridge::operator()(double Spot) const
{
    return ThePayOffPtr->operator ()(Spot);
}

#endif
```

Listing 4.14 (PayOffBridge.cpp)

```
#include<PayOffBridge.h>

PayOffBridge::PayOffBridge(const PayOffBridge& original)
{
```

```
    ThePayOffPtr = original.ThePayOffPtr->clone();
}


PayOffBridge::PayOffBridge(const PayOff& innerPayOff)
{
    ThePayOffPtr = innerPayOff.clone();
}


PayOffBridge::~PayOffBridge()
{
    delete ThePayOffPtr;
}


PayOffBridge& PayOffBridge::operator=
        (const PayOffBridge& original)
{
    if (this != &original)
    {
        delete ThePayOffPtr;
        ThePayOffPtr = original.ThePayOffPtr->clone();
    }

    return *this;
}
```

We test the new class in `VanillaMain3.cpp`. This file is almost identical to `VanillaMain2.cpp` so we do not present it here but it is included on the accompanying CD. It includes `Vanilla3.h` instead of `Vanilla2.h` and `SimpleMC5.h` instead of `SimpleMC4.h`, the only differences being the inclusion of `PayOff-Bridge.h`. In `Vanilla3.h`, we code the vanilla option in the way we originally wanted to! That is, we treat the pay-off as an ordinary object requiring no special treatment.

Listing 4.15 (`Vanilla3.h`)

```
#ifndef VANILLA_3_H
#define VANILLA_3_H


#include <PayOffBridge.h>


class VanillaOption
{
```

```
public:
 VanillaOption(const PayOffBridge&
                              ThePayOff_,double Expiry);

  double OptionPayOff(double Spot) const;
  double GetExpiry() const;

private:
  double Expiry;
  PayOffBridge ThePayOff;
};
#endif
```

and

Listing 4.16 (Vanilla3.cpp)

```
#include <Vanilla3.h>

VanillaOption::VanillaOption(const PayOffBridge&
                              ThePayOff_, double Expiry_)
                : ThePayOff(ThePayOff_), Expiry(Expiry_)
{
}

double VanillaOption::GetExpiry() const
{
    return Expiry;
}

double VanillaOption::OptionPayOff(double Spot) const
{
    return ThePayOff(Spot);
}
```

Everything in Vanilla3.h is totally straightforward as all the work has already been done for by the bridged class.

An interesting aspect of VanillaMain3.cpp is that we do not have to change the lines

```
    PayOffCall thePayOff(Strike);

    VanillaOption theOption(thePayOff, Expiry);
```

Although the `VanillaOption` constructor expects an argument of type `PayOffBridge`, it happily accepts the argument of type `PayOffCall`. The reason is that there is a constructor for `PayOffBridge` which takes in an object of type `PayOff&`. The compiler automatically accepts the inherited class object as a substitute for the base class object, and then silently converts it for us into the `PayoffBridge` object which is then passed to the `VanillaOption` constructor.

4.6 Beware of new

We have presented the bridge as a solution to the problem of developing an open-closed pay-off class. It is a very good solution and we will use this approach again and again. However, there is a downside which it is important to be aware of:

> `new` is slow.

Everytime we copy a bridged object, we are implicitly using the `new` command. So if our code involves a lot of passing around of bridged objects, we had best be sure that we do not copy them unnecessarily, or we will spend all our time calling the `new` function. One easy way to reduce the amount of copying is to always pass parameters by reference.

Why is `new` slow? To understand this, we first have to understand what it actually does. When we create variables normally, that is when not using `new`, the compiler gets them from an area of memory known as the stack. The important point is that these variables are always destroyed in reverse order from their creation (which is why the memory area is called the *stack*). Each variable as declared is added to the top of the stack, and when destroyed is removed from the top; in fact, all that happens is that the code remembers a different point as the top.

This is very easy to do but really depends rather heavily on the fact that the order of creation is the reverse of the order of destruction. Thus the variable to be destroyed is always the variable at the top of the stack.

Suppose we want to destroy variables in a different order, which is what we are really doing when we use `new`; remember that a variable created by `new` persists until the coder tells it to disappear. If we were using the stack we would have to scan down to the point where the variable was stored. We would then need to move all the variables further up the stack down a bit to cover up the gap caused by the release of memory. Clearly, this could be very time-consuming.

The compiler therefore does not use the stack for `new` but instead uses a different area of memory known as the *heap*. For this area of memory, the code keeps track of which pieces are in use and which pieces are not. Everytime `new` is called, the compiler has to find an empty piece of memory which is large enough and mark the memory as being in use. When `delete` is called, the code marks the memory as being free again.

The upshot of all this is that calling `new` involves a lot of behind the scenes work which you would probably rather not have to think about.

One solution to the time-consumption of `new` is to rewrite it. This is allowed by the standard. It is, perhaps, a little surprising that one can rewrite `new` to be more efficient than a good compiler. The reason that this is possible is that the coder has a better idea of what objects will be needed during his program's run, and possibly in what order they will be created. This allows the programmer to make extra assumptions not available to the compiler, and thus to speed things up. For example, if the programmer knows that a lot of `PayOffCall` objects will be used, he could section off an area just for `PayOffCall` objects and allocate them and deallocate them as necessary.

We shall not further explore how to rewrite `new`, but finish with an admonition that it should never be used explicitly nor implicitly within time-critical loops.

4.7 A parameters class

We know that eventually the evil boss is going to demand a variable parameters Monte Carlo routine. Let's apply the ideas of this chapter to developing a `Parameters` class which allows easy extension to variable parameters without actually putting them in. Note the crucial distinction between including variable parameters, and redesigning so that it would be easy to add them if one so desired.

What should a parameters class do? We want it to be able to store parameters such as volatility or interest rates or, in a more general set-up, jump intensity. What information will we want to get out from the class? When implementing a financial model, we never actually need the instantaneous value of parameter: it is always the integral or the integral of the square that is important.

For example, in our simple Monte Carlo model, we need the square integral of the volatility from time zero to expiry, and the integral of r over the same interval. Similarly, for jump intensity it is the integral over a time interval that is important.

Our parameters class should therefore be able to tell us the integral or integral square over any time interval, and nothing else.

What sort of differing parameters might we want? It is important in object-oriented programming to think to the future, and therefore think about not just what we want now but what we might want in the future. The simplest possible sort of parameter is a constant and our class should certainly able to encompass that. Some other obvious sorts of parameters are a polynomial, a piece-wise constant function, and an exponentially decaying function. For all of these, finding the integral and square integral over definite intervals is straightforward.

We employ the bridge design. We therefore define a class `Parameters` that handles the interaction without the outside world, and the memory handling. Its only data member is a pointer to an abstract class, `ParametersInner`, which defines the interface that the concrete classes we will eventually implement must fulfill. We also include the `clone` method, as well as `Integral` and `IntegralSquare` methods, so the wrapper class, `Parameters`, can handle the copying. We then inherit classes such as `ParametersConstant` from `ParametersInner`. As with the payoff class, additional classes can then be added in separate files, and their addition will not require any recompilation or changes in routines that use the `Parameters` class.

We give a possible implementation:

Listing 4.17 (`Parameters.h`)

```
#ifndef PARAMETERS_H
#define PARAMETERS_H

class ParametersInner
{

public:
  ParametersInner(){}

    virtual ParametersInner* clone() const=0;
    virtual double Integral(double time1,
                            double time2) const=0;
    virtual double IntegralSquare(double time1,
                                  double time2) const=0;

private:
};

class Parameters
{

public:
    Parameters(const ParametersInner& innerObject);
    Parameters(const Parameters& original);
    Parameters& operator=(const Parameters& original);
    virtual ~Parameters();
```

```
    inline double Integral(double time1, double time2) const;
    inline double IntegralSquare(double time1,
                                 double time2) const;

    double Mean(double time1, double time2) const;
    double RootMeanSquare(double time1, double time2) const;

private:
    ParametersInner* InnerObjectPtr;
};

inline double Parameters::Integral(double time1,
                                   double time2) const
{
    return InnerObjectPtr->Integral(time1,time2);
}

inline double Parameters::IntegralSquare(double time1,
                                         double time2) const
{
    return InnerObjectPtr->IntegralSquare(time1,time2);
}

class ParametersConstant : public ParametersInner
{
public:
    ParametersConstant(double constant);

    virtual ParametersInner* clone() const;
    virtual double Integral(double time1, double time2) const;
    virtual double IntegralSquare(double time1,
                                  double time2) const;

private:
    double Constant;
    double ConstantSquare;
};
#endif
```

and the source file

Listing 4.18 (Parameters.cpp)

```cpp
#include <Parameters.h>

Parameters::Parameters(const ParametersInner& innerObject)
{
    InnerObjectPtr = innerObject.clone();
}

Parameters::Parameters(const Parameters& original)
{
    InnerObjectPtr = original.InnerObjectPtr->clone();
}

Parameters& Parameters::operator=(const Parameters& original)
{
    if (this != &original)
    {
        delete InnerObjectPtr;
        InnerObjectPtr = original.InnerObjectPtr->clone();
    }
    return *this;
}

Parameters::~Parameters()
{
    delete InnerObjectPtr;
}

double Parameters::Mean(double time1, double time2) const
{
    double total = Integral(time1,time2);
    return total/(time2-time1);
}

double Parameters::RootMeanSquare(double time1,
                                  double time2) const
{
    double total = IntegralSquare(time1,time2);
    return total/(time2-time1);
```

```
}


ParametersConstant::ParametersConstant(double constant)
{
    Constant = constant;
    ConstantSquare = Constant*Constant;
}


ParametersInner* ParametersConstant::clone() const
{
    return new ParametersConstant(*this);
}


double ParametersConstant::Integral(double time1,
                                    double time2) const
{
    return (time2-time1)*Constant;
}


double ParametersConstant::IntegralSquare(double time1,
                                          double time2) const
{
    return (time2-time1)*ConstantSquare;
}
```

We have added a couple of useful, if unnecessary, methods to the `Parameters` class. The `Mean` method returns the average value of the parameter between two times, and is implemented by calling the `Integral` method which does all the work. The `RMS` method returns the root-mean-square of the parameter between two times. As it is the root-mean-square of the volatility that is appropriate to use in the Black–Scholes formula when volatility is variable, this comes in useful.

The rest of the code works in the same way as for the `PayOff` class. We give one inherited class, the simplest possible one, `ParametersConstant`, which enacts a constant parameter. As well as storing the value of the constant, we also store its square in order to minimize time spent computing the square integral. In this case, the saving is rather trivial, but the principle that time can be saved by computing values, which may be needed repeatedly, once and for all in the constructor, is worth remembering.

In `SimpleMC6`, we give a modified implementation of the simple Monte Carlo which uses the new classes. First though we need

Listing 4.19 (SimpleMC6.h)

```
#ifndef SIMPLEMC6_H
#define SIMPLEMC6_H

#include <Vanilla3.h>
#include <Parameters.h>

double SimpleMonteCarlo4(const VanillaOption& TheOption,
                         double Spot,
                         const Parameters& Vol,
                         const Parameters& r,
                         unsigned long NumberOfPaths);


#endif
```

Listing 4.20 (SimpleMC6.cpp)

```
#include<SimpleMC6.h>
#include <Random1.h>
#include <cmath>

// the basic math functions should be in namespace
// std but aren't in VCPP6
#if !defined(_MSC_VER)
using namespace std;
#endif

double SimpleMonteCarlo4(const VanillaOption& TheOption,
                         double Spot,
                         const Parameters& Vol,
                         const Parameters& r,
                         unsigned long NumberOfPaths)
{

    double Expiry = TheOption.GetExpiry();
    double variance = Vol.IntegralSquare(0,Expiry);
    double rootVariance = sqrt(variance);
```

```
    double itoCorrection = -0.5*variance;

    double movedSpot = Spot*exp(r.Integral(0,Expiry) +
                                    itoCorrection);

    double thisSpot;

    double runningSum=0;

    for (unsigned long i=0; i < NumberOfPaths; i++)
    {
        double thisGaussian = GetOneGaussianByBoxMuller();

        thisSpot = movedSpot*exp( rootVariance*thisGaussian);

        double thisPayOff = TheOption.OptionPayOff(thisSpot);

        runningSum += thisPayOff;
    }

    double mean = runningSum / NumberOfPaths;

    mean *= exp(-r.Integral(0,Expiry));

    return mean;
}
```

with the obvious corresponding changes in the header file. The main difference from the previous version is that instead of computing Vol*Vol*Expiry and r*Expiry, the IntegralSquare and Integral methods of the Parameters class are called. Note that by forcing us to go via the methods of the Parameters class, the new design makes the code more comprehensible. We know immediately from looking at it that the integral of *r* is used, as is the square-integral of the volatility.

In VanillaMain4.cpp, we adapt our main program to use the Parameters class. The main difference from the previous version is that we create a ParametersConstant object from the inputs which is then passed into the Monte Carlo routine. Note that we do not have to create the Parameters object explicitly; the conversion is done implicitly by the compiler. The relevant portion of code is

```
ParametersConstant VolParam(Vol);
ParametersConstant rParam(r);

double result = SimpleMonteCarlo4(theOption,
                                 Spot,
                                 VolParam,
                                 rParam,
                                 NumberOfPaths);
```

4.8 Key points

- Cloning gives us a method of implementing a virtual copy constructor.
- The rule of three says that if we need any one of copy constructor, destructor and assignment operator then we need all three.
- We can use a wrapper class to hide all the memory handling, allowing us to treat a polymorphic object just like any other object.
- The bridge pattern allows us to separate interface and implementation, enabling us to vary the two independently.
- The new command is slow.
- We have to be careful to ensure that self-assignment does not cause crashes.

4.9 Exercises

Exercise 4.1 Test how fast new is on your computer and compiler. Do this by using it to allocate an array of doubles, ten thousand times. See how the speed varies with array size. If you have more than one compiler see how they compare. Note that you can time things using the clock() function.

Exercise 4.2
Find out about an auto_ptr. Observe that it cannot be copied in the usual sense of copying. Show that a class with an auto_ptr data member will need a copy constructor but not a destructor.

Exercise 4.3
Implement a piecewise-constant parameters class.

5

Strategies, decoration and statistics

5.1 Differing outputs

Our Monte Carlo routine is now easily extendible to handle any pay-off and time-dependent parameters. However, there are plenty of valid criticisms that could still be made. One thing that is definitely lacking is the absence of any indication of the simulation's convergence. We could make the routine return standard error, or a convergence table, or simply have it return the value for every single path and analyze the results elsewhere.

As we are trying to develop an object-oriented routine, we make the statistics gatherer an input. Thus the Monte Carlo routine will take in a statistics gatherer object, store the results in it and the statistics gatherer will then output the statistics as required. This technique of using an auxiliary class to decide how part of an algorithm is implemented is sometimes called the *strategy pattern*.

5.2 Designing a statistics gatherer

We want our statistics gatherer to be reusable; there are plenty of circumstances where such a routine might be useful. For example, we might have many other Monte Carlo routines such as an exotics pricer or a BGM pricer for interest-rate derivatives. Also, if we are developing a risk system, we might be more interested in the ninety-fifth percentile, or in the conditional expected shortfall, than in the mean or variance.

What should the routine do? It must have two principal methods. The first should take in data for each path. The second must output the desired statistics.

Since we do not wish to specify what sort of statistics are being gathered in advance, we proceed via an abstract base class using virtual methods, just as we did for the `PayOff` and `Parameters` classes. However, as most of the time we will not need to copy these objects we do not bother with the bridge but work with the base class by reference.

We have to decide the precise interface for our two principal methods. They will be pure virtual functions declared in the base class and defined in the concrete inherited class. Our first method, DumpOneResult, we make quite simple: it takes in a double and returns nothing. Note that it is not a const method, since by its very nature it must update the statistics stored inside the object. Note that we have not allowed the possibility of dumping more than one value per path, which could be argued to be a defect. The object will store what it needs to in order to compute the statistics desired and no more. So if statistics gatherer's job is to compute the mean, then it need only store the running sum and the number of paths. However, for a more complicated statistic we might need the value for every path to be stored.

Our second method, which will indeed return the results, requires a little more thought. We have to decide what sort of object to return the results in. Another issue is whether it should be possible to ask for statistics en route or ought we be able to call the method for returning results only once.

With regard to the form in which to return results, we opt for a vector of vectors. This will allow us to easily return a table if we so desire. Whilst this would not be a great way to implement a matrix-class if we were doing linear algebra, the return statistics are not a matrix, just a table and this is sufficient for our purposes.

We opt to allow the return statistics method to be called many times and therefore name it GetResultsSoFar. This will cost us little (possibly nothing), and will be more robust than an object that crashes if the get results method is called twice. We make it a const method as it should not change the state of the object in any substantive way: this enforces the rule that the method can be called many times.

Listing 5.1 (MCStatistics.h)

```cpp
#ifndef STATISTICS_H
#define STATISTICS_H

#include <vector>

class StatisticsMC
{
public:
    StatisticsMC(){}

    virtual void DumpOneResult(double result)=0;
    virtual std::vector<std::vector<double> >
                    GetResultsSoFar() const=0;
```

```
    virtual StatisticsMC* clone() const=0;
    virtual ~StatisticsMC(){}

private:
};

class StatisticsMean : public StatisticsMC
{

public:
    StatisticsMean();
    virtual void DumpOneResult(double result);
    virtual std::vector<std::vector<double> >
                        GetResultsSoFar() const;
    virtual StatisticsMC* clone() const;

private:
    double RunningSum;
    unsigned long PathsDone;
};
#endif
```

Our abstract base class is `StatisticsMC`. It has the pure virtual functions `Dump-OneResult` and `GetResultsSoFar`. We include the `clone` method to allow for the possibility of virtual copy construction. We also make the destructor virtual, as any cloned objects will likely need to be destroyed via pointers to the base class which will not know their type, as usual. The base class does nothing but the important task of defining an interface.

We give a very simple inherited class `StatisticsMean`, which returns the mean of the simulation, just as our routine previously did. The source code is included in `MCStatistics.cpp`.

Listing 5.2 (`MCStatistics.cpp`)

```
#include<MCStatistics.h>
using namespace std;

StatisticsMean::StatisticsMean()
    :
    RunningSum(0.0), PathsDone(0UL)
{
```

```
}

void StatisticsMean::DumpOneResult(double result)
{
    PathsDone++;
    RunningSum += result;
}

vector<vector<double> >
        StatisticsMean::GetResultsSoFar() const
{
    vector<vector<double> > Results(1);

    Results[0].resize(1);
    Results[0][0] = RunningSum / PathsDone;

    return Results;
}

StatisticsMC* StatisticsMean::clone() const
{
    return new StatisticsMean(*this);
}
```

Note that whilst we write the `DumpOneResult` method to be efficient since it will be called in every iteration of the loop, we do not worry about efficiency for `GetResultsSoFar`, as it will generally be called only once per simulation.

5.3 Using the statistics gatherer

Having written a class for gathering statistics, we now need to modify our Monte Carlo routine to make use of it. We do this in the function `SimpleMonteCarlo5` which is declared in `SimpleMC7.h`.

Listing 5.3 (`SimpleMC7.h`)

```
#ifndef SIMPLEMC7_H
#define SIMPLEMC7_H

#include <Vanilla3.h>
#include <Parameters.h>
```

```
#include <MCStatistics.h>

void SimpleMonteCarlo5(const VanillaOption& TheOption,
                       double Spot,
                       const Parameters& Vol,
                       const Parameters& r,
                       unsigned long NumberOfPaths,
                       StatisticsMC& gatherer);
#endif
```

Note that the `StatisticsMC` object is passed in by reference and is not `const`. This is crucial as we want the object passed in to gather the information given to it and for this information to be available after the function has returned. If we had passed by value then the object outside would not change, and all the results would disappear at the end of the function which we emphatically do not want. If the object was `const`, then it would not be possible to put any new data into it which would be useless. Previously, our routine returned a `double`; now it is `void` as all the data to be returned is inside the object `gatherer`.

We define the function in `SimpleMC7.cpp`:

Listing 5.4 (`SimpleMC7.cpp`)

```
#include <SimpleMC7.h>
#include <Random1.h>
#include <cmath>
// the basic math functions should be
// in namespace std but aren't in VCPP6
#if !defined(_MSC_VER)
using namespace std;
#endif

void SimpleMonteCarlo5(const VanillaOption& TheOption,
                       double Spot,
                       const Parameters& Vol,
                       const Parameters& r,
                       unsigned long NumberOfPaths,
                       StatisticsMC& gatherer)
{
    double Expiry = TheOption.GetExpiry();
    double variance = Vol.IntegralSquare(0,Expiry);
    double rootVariance = sqrt(variance);
```

```
double itoCorrection = -0.5*variance;
double movedSpot =
    Spot*exp(r.Integral(0,Expiry)+itoCorrection);
double thisSpot;
double discounting = exp(-r.Integral(0,Expiry));

for (unsigned long i=0; i < NumberOfPaths; i++)
{
    double thisGaussian = GetOneGaussianByBoxMuller();
    thisSpot = movedSpot*exp( rootVariance*thisGaussian);
    double thisPayOff = TheOption.OptionPayOff(thisSpot);
    gatherer.DumpOneResult(thisPayOff*discounting);
}
return;
}
```

Our routine appears simpler than `SimpleMonteCarlo4`; all the work previously spent accounting the results is now sublimated into the line on which we call `.DumpOneResult`. Of course, the code has not disappeared; it has simply moved into a different file. Thus the strategy pattern gives us a readability benefit as well as flexibility.

We illustrate how the gatherer might be called in `StatsMain1.cpp`:

Listing 5.5 (`StatsMain1.cpp`)

```
/*
 uses source files
    MCStatistics.cpp,
    Parameters.cpp,
    PayOff3.cpp,
    PayOffBridge.cpp,
    Random1.cpp,
    SimpleMC7.cpp,
    Vanilla3.cpp,
*/
#include<SimpleMC7.h>
#include<iostream>
using namespace std;
#include<Vanilla3.h>
#include<MCStatistics.h>
```

```
int main()
{
    double Expiry;
    double Strike;
    double Spot;
    double Vol;
    double r;
    unsigned long NumberOfPaths;

    cout << "\nEnter expiry\n";
    cin >> Expiry;

    cout << "\nStrike\n";
    cin >> Strike;

    cout << "\nEnter spot\n";
    cin >> Spot;

    cout << "\nEnter vol\n";
    cin >> Vol;

    cout << "\nr\n";
    cin >> r;

    cout << "\nNumber of paths\n";
    cin >> NumberOfPaths;

    PayOffCall thePayOff(Strike);

    VanillaOption theOption(thePayOff, Expiry);

    ParametersConstant VolParam(Vol);
    ParametersConstant rParam(r);

    StatisticsMean gatherer;

    SimpleMonteCarlo5(theOption,
                      Spot,
                      VolParam,
                      rParam,
```

```
                    NumberOfPaths,
                    gatherer);

vector<vector<double> > results = gatherer.GetResultsSoFar();

    cout <<"\nFor the call price the results are \n";

    for (unsigned long i=0; i < results.size(); i++)
    {
        for (unsigned long j=0; j < results[i].size(); j++)
            cout << results[i][j] << " ";

        cout << "\n";
    }
    double tmp;
    cin >> tmp;

    return 0;
}
```

Our output of the results is now a bit more complicated in that we have to loop over the `vector` of `vectors` in order to display the results. Of course, for the particular class we have defined, only one number is returned so this is not strictly necessary. However, by writing it for the most general sort of return statement from a statistics gatherer, we produce more robust code. After all, the object has contracted to return a `vector` of `vectors`, and we should code in accordance with this contract, and make no extra assumptions.

5.4 Templates and wrappers

We have created a class hierarchy for gathering statistics. This hierarchy includes a virtual constructor, `clone`, so we can copy these objects without knowing their type. However, if we wish to start copying and storing the objects then we have a slight issue in that, just as with our `PayOff` class, this will have to be done manually unless we write an extra wrapper class to do it for us. In the next section, we give an example where this copying will be necessary, so we do need to provide a wrapper class.

It becomes clear at this point that we will need to write these wrapper classes repeatedly in a very similar way. We therefore present a templatized solution. The name of the base class to be wrapped is taken as an argument to the wrapper class and it can then be used for any class which provides a `clone` method.

Our `Wrapper` class provides various functionalities which are intended to make it act like a pointer to a single object but with added responsibilities. The added responsibilities are that the pointer is both responsible for and owns the object pointed to at all times. Thus if we copy the `Wrapper` object, the pointed-to object is also copied, so that each `Wrapper` object has its own copy of the pointed-to object. When the `Wrapper` object ceases to exist because of going out of scope, or being `deleted`, the pointed-to object is automatically `deleted` as well.

If we set one `Wrapper` object equal to another, then the object previously pointed to must be `deleted`, and then a copy of the new object must be created so each `Wrapper` still owns precisely one object.

In addition, it must be possible to dereference the `Wrapper` to obtain the underlying object. In other words, if you put `*mywrapper` then you should obtain the object pointed to by `mywrapper`. We do this by overloading the `operator*()` and just make it return the dereferenced inner pointer.

We also want to be able to access the methods of the inner object. Whilst this can always be done by putting `(*mywrapper).theMethod()`, it is a lot less elegant than being able to type `myWrapper->theMethod()`, which is the normal code for an ordinary pointer. We provide this functionality by overloading `operator->()`.

We provide the relevant code in `Wrapper.h`.

Listing 5.6 (`Wrapper.h`)

```
#ifndef WRAPPER_H
#define WRAPPER_H

template< class T>
class Wrapper
{
public:
    Wrapper()
    { DataPtr =0;}

    Wrapper(const T& inner)
    {
        DataPtr = inner.clone();
    }

    ~Wrapper()
    {
        if (DataPtr !=0)
```

```
        delete DataPtr;
}

Wrapper(const Wrapper<T>& original)
{
    if (original.DataPtr !=0)
        DataPtr = original.DataPtr->clone();
    else
        DataPtr=0;
}

Wrapper& operator=(const Wrapper<T>& original)
{
    if (this != &original)
    {
        if (DataPtr!=0)
            delete DataPtr;

        DataPtr = (original.DataPtr !=0) ?
            original.DataPtr->clone() : 0;
    }

    return *this;
}

T& operator*()
{
    return *DataPtr;
}

const T& operator*() const
{
    return *DataPtr;
}

const T* const operator->() const
{
    return DataPtr;
}
```

```
    T* operator->()
    {
        return DataPtr;
    }

private:
    T* DataPtr;
};
#endif
```

We start before each declaration with the command `template<class T>` to let the compiler know we are writing template code. The class T will be specified elsewhere. The compiler will produce one copy of the code for each different sort of T that is used. Thus if we declare `Wrapper<MCStatistics>` `TheStatsGatherer;`, the compiler will then proceed to create the code by substituting `MCStatistics` for T everywhere, and then compile it. This has some side effects: the first is that all the code for the `Wrapper` template is in the header file – there is no `wrapper.cpp` file. The second is that if we use the `Wrapper` class many times, we have to compile a lot more code than we might actually expect. Whilst this is not really an issue for this class, it could be one for a complicated class, where we might end up with rather slow compile times and a much larger than expected executable. There are some other effects and we will return to this topic in Section 9.6.

We provide the `Wrapper` class with a default constructor. This means that it is possible to have a `Wrapper` object which points to nothing. If we did not then a declaration such as

```
std::vector<Wrapper<MCStatistics> > StatisticsGatherers(10);
```

would not compile: the constructor for `vector` would look for the default constructor for `Wrapper<MCStatistics>` in order to create the ten copies specified, and not find it. Why would we want a `vector` of `Wrappers`? We saw in Section 3.4 that we can get into trouble if we try to copy inherited objects into base class objects without a wrapper class. The same reasons apply here. We cannot declare a `vector` of base class objects as they are abstract, and even if they were not, they would be the wrong size. We therefore have to store pointers, references or wrappers, and wrappers are the easiest option; they take care of all the memory handling for us.

Given that a `Wrapper` object can point to nothing, we have to be able to take this into account when writing the class's methods. We indicate that the object points to nothing by setting the pointer to zero. When carrying out copying and assignment, we then have to take care of this special case.

We provide two different versions of the dereferencing operator *, as it should be possible to dereference both `const` and non-`const` objects. As one would expect, the `const` version returns a `const` object and the non-`const` version does not. We have declared the two operators `inline` to ensure that there is no performance overhead induced by going via a wrapper.

Similarly, we declare the `operator->` to have both `const` and non-`const` versions. The syntax here is a little strange in that all the operator does is return the pointer. However, there are special rules for overloading `->` which ensure that any method following `->` is correctly invoked for the pointer returned.

5.5 A convergence table

If we use a statistics gatherer and run the simulation it will tell us the relevant statistics for the entire simulation. However, it does not necessarily give us a feel for how well the simulation has converged. One standard method of checking the convergence is to examine the standard error of the simulation; that is measure the sample standard deviation and divide by the square root of the number of paths. If one is using low-discrepancy numbers this measure does not take account of their special properties and, in fact, it predicts the same error as for a pseudo-random simulation (see for example [10]). Which we can expect to be too large. (Else why use low-discrepancy numbers?)

One alternative method is therefore to use a convergence table. Rather than returning statistics for the entire simulation, we instead return them for every power of two to get an idea of how the numbers are varying. We could just write a class directly to return such a table for the mean, but since we might want to do this for any statistic, we do it in a reusable fashion.

Our class must contain a statistics gatherer in order to decide for which statistics to create a convergence table. On the other hand, it must implement the same interface as all the other statistics gatherers so we can plug it into the same simulations. We therefore define a class `ConvergenceTable` which is inherited from `MCStatistics`, and has a wrapper of an `MCStatistics` object as a data member.

The fact that the class is inherited from `MCStatistics` guarantees that from the outside it looks just like any other statistics-gatherer object. The difference on the inside is that we can make the data member refer to any kind of statistics gatherer we like, and so we have a convergence table for any statistic for which a statistics gatherer has been written. We give the implementation in `ConvergenceTable.h` and `ConvergenceTable.cpp`.

Listing 5.7 (`ConvergenceTable.h`)

```
#ifndef CONVERGENCE_TABLE_H
#define CONVERGENCE_TABLE_H
#include <MCStatistics.h>
#include <wrapper.h>

class ConvergenceTable : public StatisticsMC
{
public:
    ConvergenceTable(const Wrapper<StatisticsMC>& Inner_);

    virtual StatisticsMC* clone() const;
    virtual void DumpOneResult(double result);
    virtual std::vector<std::vector<double> >
                        GetResultsSoFar() const;

private:
    Wrapper<StatisticsMC> Inner;
    std::vector<std::vector<double> > ResultsSoFar;
    unsigned long StoppingPoint;
    unsigned long PathsDone;
};

#endif
```

Listing 5.8 (`ConvergenceTable.cpp`)

```
#include<ConvergenceTable.h>

ConvergenceTable::ConvergenceTable(const
                        Wrapper<StatisticsMC>& Inner_)
: Inner(Inner_)
{
    StoppingPoint=2;
    PathsDone=0;
}

StatisticsMC* ConvergenceTable::clone() const
{
    return new ConvergenceTable(*this);
}
```

```
void ConvergenceTable::DumpOneResult(double result)
{
    Inner->DumpOneResult(result);
    ++PathsDone;

    if (PathsDone == StoppingPoint)
    {
        StoppingPoint*=2;
        std::vector<std::vector<double> >
            thisResult(Inner->GetResultsSoFar());

        for (unsigned long i=0; i < thisResult.size(); i++)
        {
            thisResult[i].push_back(PathsDone);
            ResultsSoFar.push_back(thisResult[i]);
        }
    }

    return;
}


std::vector<std::vector<double> >
            ConvergenceTable::GetResultsSoFar() const
{
    std::vector<std::vector<double> > tmp(ResultsSoFar);

    if (PathsDone*2 != StoppingPoint)
    {
        std::vector<std::vector<double> >
            thisResult(Inner->GetResultsSoFar());

        for (unsigned long i=0; i < thisResult.size(); i++)
        {
            thisResult[i].push_back(PathsDone);
            tmp.push_back(thisResult[i]);
        }
    }
    return tmp;
}
```

Note that we do not write a copy constructor, destructor or assignment operator as the class itself does no dynamic memory allocation. Dynamic memory allocation does occur inside the class but it is all handled automatically by the `Wrapper` template class.

The class does not do a huge amount; every result passed in is passed to the inner class. When we reach a point where the number of paths done is a multiple of two, the inner class's `GetResults()` method is called, and the results stored with the number of paths done so far added in. When the class's own `GetResults()` methods is called, it calls the inner class's method one more time if necessary and then spits out all the stored results.

In `StatsMain2.cpp`, we illustrate how the routine might be called:

Listing 5.9

```
StatisticsMean gatherer;
ConvergenceTable gathererTwo(gatherer);

SimpleMonteCarlo5(theOption,
                  Spot,
                  VolParam,
                  rParam,
                  NumberOfPaths,
                  gathererTwo);

vector<vector<double> > results =
                            gathererTwo.GetResultsSoFar();
```

First create a `StatisticsMean` object: then pass it into a `ConvergenceTable` object, `gatherTwo`. Note the constructor takes a `Wrapper<MCStatistics>` object but the compiler happily does this conversion for us. We then pass the new gatherer into `SimpleMonteCarlo5` which has not required any changes. We have also not made any changes to either of the `MCStatistics` files.

5.6 Decoration

The technique of the last section is an example of a standard design pattern called the *decorator pattern*. We have added functionality to a class without changing the interface. This process is called *decoration*. The most important point is that, since the decorated class has the same interface as the undecorated class, any decoration which can be applied to the original class can also be applied to the decorated class.

We can therefore decorate as many times as we wish. It would be syntactically legal (but not useful), for example, to have a convergence table of convergence tables. We will more often wish to decorate several times but in differing manners.

How else might we want to decorate? If we have a stream of numbers defining a time series, we often want the statistics of the successive increments instead of the numbers themselves. A decorator class could therefore do this differencing and pass the difference into the inner class.

We might want more than one statistic for a given set of numbers; rather than writing one class to gather many statistics, we could write a decorator class which contains a vector of statistics gatherers and passes the gathered value to each one individually. The `GetResults()` method would then garner the results from each of the inner gatherers and collate them.

We can also apply these decoration ideas to the `Parameters` class. We could define a class that takes the linear multiple of an inner `Parameters` object for example. This class would simple multiply the integral by a given constant, and the square integral by its square.

5.7 Key points

In this chapter, we have seen that we can allow the user to specify aspects of how an algorithm works by making part of the algorithm be carried out in an inputted class. We have also examined the techniques of decoration and templatization.

- Routines can be made more flexible by using the strategy pattern.
- Making part of an algorithm be implemented by an inputted class is called the strategy pattern.
- For code that is very similar across many different classes, we can use templates to save time in rewriting.
- If we want containers of polymorphic objects, we must use wrappers or pointers.
- Decoration is the technique of adding functionality by placing a class around a class which has the same interface; i.e. the outer class is inherited from the same base class.
- A class can be decorated several times.

5.8 Exercises

Exercise 5.1 Write a statistics gathering class that computes the first four moments of a sample.

Exercise 5.2 Write a statistics gathering class that computes the value at risk of a sample.

Exercise 5.3 Write a statistics gathering class that allows the computation of several statistics via inputted classes.

Exercise 5.4 Use the strategy pattern to allow the user to specify termination conditions for the Monte Carlo, e.g., time spent or paths done.

Exercise 5.5 Write a terminator class that causes termination when either of two inner terminator classes specifies termination.

Exercise 5.6 * Write a template class that implements a reference counted wrapper. This will be similar to the wrapper class but instead of making a clone of the inner object when the wrapper is copied, an internal counter is increased and the inner object is shared. When a copy is destroyed, the inner counter is decremented. When the inner counter reaches zero, the object is destroyed. Note that both the counter and the inner object will be shared across copies of the object. (This exercise is harder than most in this book.)

6

A random numbers class

6.1 Why?

So far, we have been using the inbuilt random number generator, rand. In this chapter, we look at how we might implement a class to encapsulate random number generation. There are a number of reasons we might wish to do this.

rand is implementation dependent. The standard specifies certain properties of rand and gives an example of how it could be implemented but it does not actually specify the details. This has important consequences for us. The first is simply that we cannot expect any consistency across compilers. If we decide to test our code by running it on multiple platforms, we can expect to obtain differing streams of random numbers and whilst our Monte Carlo simulations should still converge to the same number, this is a lot weaker than having every single random draw matching. Thus our code becomes harder to test. A second issue is that we do not know how good the compiler's implementation is. Either we have to get hold of technical documents for every compiler we use and make sure that the implementors have done a good job, or we have to run a number of statistical tests to ensure that rand is up to the job. Note that for most simulations we will actually need many random draws for each path, and so it is not enough for us to check that single draws do a good job of simulating the uniform distribution; instead we need a large number of successive draws to do a good job of filling out the unit hypercube, which is much tougher.

rand is not predictable. A crucial aspect of running Monte Carlo simulations is that they must be reproducible. If we run the same simulation twice we want to obtain precisely the same random numbers. We can achieve this with rand by using the srand command to set the seed which will guarantee the same number stream from rand every time. The problem is that the seed is a global variable. This means that calling rand in different parts of the program will cause totally unrelated pieces of code to affect each other's operation. We therefore want to be

83

able to insulate the random number stream used by a particular simulation from the rest of the program.

Another advantage of using a class is that we can decorate it. For example, suppose we wish to use anti-thetic sampling. We could write a decorator class that does anti-thetic sampling. This can then be combined with any random number generator we have written, and plugged into the Monte Carlo simulator, with no changes to the simulator class. If we used `rand` directly we would have to fiddle with the guts of the simulator class. Similarly, if we wish to carry out moment matching we could use a decorator class and then plug the decorated class into the simulator.

A further reason is that we might decide not to use pseudo-random (i.e. random) numbers but low-discrepancy numbers instead. Low-discrepancy numbers (sometimes called quasi-random numbers) are sequences of numbers designed to do a good job of filling out space. They are therefore anything but random. However, they have the right statistical properties to guarantee that simulations converge to the correct answer. Their space-filling properties mean they make simulations converge faster. If we are using a random number class, we could replace this class with a generator for low-discrepancy numbers without changing the interior of our code.

6.2 Design considerations

As we wish the possibility of having many random number generators and we want to be able to add new ones later on without recoding, we use an abstract base class to specify an interface. Each individual generator will then be inherited from it. In order to specify the interface, we have to identify what we want from any random number class.

Generally, when working with any Monte Carlo simulation, the simulation will have a dimensionality which is the number of random draws needed to simulate one path. This number is equal to the number of variables of integration in the underlying integral which we are trying to approximate. It is generally cleaner therefore to obtain all the draws necessary for a path in one go. This has the added advantage that a random number generator can protest (i.e. throw an error) if it is being used beyond its dimensional specification. Additionally, when working with low-discrepancy numbers it is essential that the generator know the dimensionality as the generator has to be set up specifically for each choice of dimension.

This means that we need methods to set the dimensionality, and to obtain an array of uniforms of size dimensionality from the generator. We also provide a method that states the dimensionality.

For financial applications, we will want standard Gaussian draws more often than uniforms so we will want a method of obtaining them instead. In fact, we can separate out the creation of the uniforms and their conversion into Gaussians. The conversion into Gaussians can therefore be done in a generator-independent fashion and this means that it can be implemented as a method of the base class which calls the virtual method that creates the uniform draws.

What else might we want? For many applications, it is necessary to generate the same stream of random numbers twice. For example, if we wish to compute Greeks by bumping parameters, the error is much smaller if the same numbers are used twice. (See for example [10] or [12].) Or if we wish to carry out moment matching, the reuse of the same random numbers stream twice enables us to avoid storing all the numbers generated. Thus we include methods to reset the generator to its initial state, and to set the seed of the generator.

Occasionally, we wish to be sure of having a different stream of random numbers. For example, when carrying out an optimization in order to estimate an exercise strategy, we generally use one set of random numbers to optimize parameters for the strategy, and then having chosen the strategy we run a second simulation with different random numbers to estimate the price. This allows us to be sure that the optimization has not exploited the micro-structure of the random number stream. A simple way to achieve the differing streams of numbers is to make sure the generator skips a number of paths equal to the number used for the first simulation. We therefore include a method which allows us to skip paths.

Finally, we may wish to copy a random number generator for which we do not know the type. We therefore include a `clone` method to enable virtual construction.

One extra issue we have to think about is in what range should a uniform lie? The uniform distribution is generally defined to be a density function on the interval $[0, 1]$ such that the probability that a draw X lies in an interval of length α is α. The subtlety lies in whether we allow the values 0 and 1 to be taken. Since taking either value is a probability zero event allowing or disallowing either value will not effect the statistical properties of our simulation, but they can have practical effects. For example, if we elect to convert the uniforms into Gaussians by using the inverse cumulative normal function (which we will) then the numbers 0 and 1 cause us difficulties since the inverse cumulative normal function naturally maps them to $-\infty$ and $+\infty$. To avoid these difficulties, we therefore require that our uniform variates never take these values and thus lie in the open interval $(0, 1)$. The main side effect of this choice is that if we use random generators written by others then we need to check that they satisfy the same convention, and if not, adapt them appropriately.

6.3 The base class

We specify the interface via a base class as follows, Random2.h,

Listing 6.1 (Random2.h)

```
#ifndef RANDOM2_H
#define RANDOM2_H

#include <Arrays.h>

class RandomBase
{
public:
    RandomBase(unsigned long Dimensionality);

    inline unsigned long GetDimensionality() const;

    virtual RandomBase* clone() const=0;
    virtual void GetUniforms(MJArray& variates)=0;
    virtual void Skip(unsigned long numberOfPaths)=0;
    virtual void SetSeed(unsigned long Seed) =0;
    virtual void Reset()=0;

    virtual void GetGaussians(MJArray& variates);
    virtual void ResetDimensionality(unsigned long
                                        NewDimensionality);

private:
    unsigned long Dimensionality;
};

unsigned long RandomBase::GetDimensionality() const
{
    return Dimensionality;
}
#endif
```

Whilst most of the methods of RandomBase are pure virtual, three are not. The method GetGaussians transforms uniforms obtained from the GetUniforms method into standard Gaussian distributions. It does this via an approximation to the inverse cumulative normal function due to Moro, [19]. As this method only uses one uniform to produce a Gaussian and enacts precisely the definition of the Gaussian distribution it is very robust and works under all circumstances. Nevertheless, we make the method virtual to allow the possibility that for a particular generator there is another preferred conversion method. Or even to allow the possibility that the generator provides normals which are then converted into uniforms by the GetUniforms method.

The GetDimensionality method simply returns the dimensionality of the generator and there is no need for it to be virtual.

We also have the concrete virtual function ResetDimensionality. As the base class stores dimensionality, it must be told when dimensionality changes: that is the purpose of this function. However, the function is virtual because generally if dimensionality changes, the random number generator will also need to know. Suppose we have overriden this virtual function in an inherited class. Calling the method thus only calls the inherited class method and the base class method is ignored; however, we still need the base class method to be called; this has to be done by the inherited class method. The syntax to do this is to prefix the method with RandomBase::. The compiler then ignores the virtual function table and instead knows to call the method associated to the base class.

Note that we define the interface for GetUniforms and GetGaussians via a reference to an array. The reason we do this is that we do not wish to waste time copying arrays. Also remember that arrays of dynamic size generally involve dynamic memory allocation, i.e. new, and therefore are quite slow to create and to destroy. We want to minimize unnecessary operations, and by passing the return values into a pre-generated array we avoid all this. The array class used here is quite simple and given in Appendix C. We assume that the array is of sufficient size. We could check that it is big enough but that could result in substantial overhead. One solution would be to check the size only if a compiler flag was set, e.g. in debug mode.

Note that one disadvantage of this approach is that we are now bound to this array class. How could we overcome that disadvantage? One solution would be to simply pass in a pointer, and write to the memory locations pointed to. However, the use of raw pointers tends to lead to code that is hard to debug, and is therefore best avoided. Another solution is to templatize so that the array class is a template argument and the code will then work with any array class which has the requisite methods. A related solution is to use iterators. An *iterator* is a generalization of a pointer and we could templatize the code to work off any iterator. We do not explore these options here but the reader should bear them in mind if he wishes to adapt the code.

The source code for the base class is quite simple as it does not do very much:

Listing 6.2 (Random2.cpp)

```cpp
#include <Random2.h>
#include <Normals.h>
#include <cstdlib>

// the basic math functions should be in namespace
// std but aren't in VCPP6
#if !defined(_MSC_VER)
using namespace std;
#endif

void RandomBase::GetGaussians(MJArray& variates)
{
    GetUniforms(variates);

    for (unsigned long i=0; i < Dimensionality; i++)
    {
        double x=variates[i];
        variates[i] = InverseCumulativeNormal(x);
    }
}

void RandomBase::ResetDimensionality(unsigned long
                                    NewDimensionality)
{
    Dimensionality = NewDimensionality;
}

RandomBase::RandomBase(unsigned long Dimensionality_)
: Dimensionality(Dimensionality_)
{
}
```

The inverse cumulative normal function is included in the file Normals and is a piece-wise rational approximation. See Appendix B.

6.4 A linear congruential generator and the adapter pattern

We now need to actually write a random number generator. A simple method of generating random numbers is a linear congruential generator. We present a generator called by Park & Miller the *minimal standard generator*. In other words, it is a generator that provides a minimum guaranteed level of statistical accuracy. We refer the reader to [26] for further discussion of this and many other random number generators.

We present the generator in two pieces. A small inner class that develops a random generator that returns one integer (i.e., `long`) every time it is called, and a larger class that turns the output into a vector of uniforms in the format desired. We present the class definition in `ParkMiller.h`.

Listing 6.3 (`ParkMiller.h`)

```
#ifndef PARK_MILLER_H
#define PARK_MILLER_H
#include <Random2.h>

class ParkMiller
{
public:
    ParkMiller(long Seed = 1);

    long GetOneRandomInteger();
    void SetSeed(long Seed);

    static unsigned long Max();
    static unsigned long Min();

private:
    long Seed;
};

class RandomParkMiller : public RandomBase
{
public:
    RandomParkMiller(unsigned long Dimensionality,
                     unsigned long Seed=1);

    virtual RandomBase* clone() const;
    virtual void GetUniforms(MJArray& variates);
    virtual void Skip(unsigned long numberOfPaths);
    virtual void SetSeed(unsigned long Seed);
    virtual void Reset();
```

```
    virtual void ResetDimensionality(unsigned long
                                     NewDimensionality);

private:
    ParkMiller InnerGenerator;
    unsigned long InitialSeed;
    double Reciprocal;
};
#endif
```

The inner class is quite simple. It develops a sequence of uncorrelated `long`s. The seed can be set either in the constructor or via a set seed method. We give two extra methods which indicate the minimum and maximum values that the generator can give out. Such information is crucial to a user who wishes to convert the output into uniforms, as they will need to subtract the minimum and divide by the maximum minus the minimum to get a number in the interval [0, 1].

The bigger class is inherited from `RandomBase`. It has all the methods that it requires. Its main data member is a `ParkMiller` generator object. It also remembers the initial seed, and the reciprocal of the maximum value plus one, to save time then turning the output of the inner generator into uniforms.

Our pattern here is an example of the *adapter* pattern. We have a random generator which works and is effective, however its interface is not what the rest of the code expects. We therefore write a class around it which adapts its interface into what we want. Whenever we use old code or import libraries, it is rare for the interfaces to fit precisely with what we have been using, and the adapter pattern is then necessary. To use the adapter pattern simply means to use an intermediary class which transforms one interface into another. It is the coding equivalent of a plug adapter.

The implementation of these classes is straightforward. The generator relies on modular arithmetic. The basic idea is that if you repeatedly multiply a number by a large number, and then take the modulus with respect to another number, then the successive remainders are effectively random. We refer the reader to [26] for discussion of the mathematics and the choice of the constants.

Listing 6.4 (`ParkMiller.cpp`)

```
#include <ParkMiller.h>

const long a = 16807;
const long m = 2147483647;
const long q = 127773;
const long r = 2836;
```

```
ParkMiller::ParkMiller(long Seed_ ) : Seed(Seed_)
{
    if (Seed ==0)
        Seed=1;
}

void ParkMiller::SetSeed(long Seed_)
{
  Seed=Seed_;
  if (Seed ==0)
        Seed=1;
}

unsigned long ParkMiller::Max()
{
    return m-1;
}

unsigned long ParkMiller::Min()
{
    return 1;
}

long ParkMiller::GetOneRandomInteger()
{
    long k;

    k=Seed/q;

    Seed=a*(Seed-k*q)-r*k;

    if (Seed < 0)
        Seed += m;

    return Seed;
}

RandomParkMiller::RandomParkMiller(unsigned long Dimensionality,
                                   unsigned long Seed)
:    RandomBase(Dimensionality),
    InnerGenerator(Seed),
```

```
    InitialSeed(Seed)
{

    Reciprocal = 1/(1.0+InnerGenerator.Max());
}


RandomBase* RandomParkMiller::clone() const
{
    return new RandomParkMiller(*this);
}


void RandomParkMiller::GetUniforms(MJArray& variates)
{
    for (unsigned long j=0; j < GetDimensionality(); j++)
        variates[j] =
        InnerGenerator.GetOneRandomInteger()*Reciprocal;
}


void RandomParkMiller::Skip(unsigned long numberOfPaths)
{
    MJArray tmp(GetDimensionality());
    for (unsigned long j=0; j < numberOfPaths; j++)
        GetUniforms(tmp);
}


void RandomParkMiller::SetSeed(unsigned long Seed)
{
    InitialSeed = Seed;
    InnerGenerator.SetSeed(Seed);
}


void RandomParkMiller::Reset()
{
    InnerGenerator.SetSeed(InitialSeed);
}


void RandomParkMiller::ResetDimensionality(unsigned long
                                        NewDimensionality)
{
    RandomBase::ResetDimensionality(NewDimensionality);
    InnerGenerator.SetSeed(InitialSeed);
}
```

Note that we check whether the seed is zero. If it is we change it to 1. The reason is that a zero seed yields a chain of zeros. Note the advantage of a class-based implementation here. The seed is only inputted in the constructor and the set seed method, which are called only rarely, so we can put in extra tests to make sure the seed is correct with no real overhead. If the seed had to be checked every time the random number generator was called, then the overhead would be substantial indeed.

The implementation of the adapter class is quite straightforward. Note that we divide the outputs of the inner class by the maximum plus 1, and so ensure that we obtain random numbers on the open interval (0, 1) rather than the closed one; this means that we will have no trouble with the inverse cumulative normal function.

6.5 Anti-thetic sampling via decoration

A standard method of improving the convergence of Monte Carlo simulations is anti-thetic sampling. The idea is very simple, if a X is a draw from a standard Gaussian distribution so is $-X$. This means that if we draw a vector $(X_1, \ldots, X_n)$ for one path then instead of drawing a new vector for the next path we simply use $(-X_1, \ldots, -X_n)$. This method guarantees that, for any even number of paths drawn, all the odd moments of the sample of Gaussian variates drawn are zero, and in particular the mean is correct. This generally, but not always, causes simulations to converge faster. See [10] for discussion of the pros and cons of anti-thetic sampling.

We wish to implement anti-thetic sampling in such a way that it can be used with any random number generator and with any Monte Carlo simulation in such a way that we only have to implement it once. The natural way to do this is the decorator pattern. The decoration can be applied to any generator so it fulfills the first criterion, and the fact that the interface is unchanged means that we can plug the decorated class into any socket which the original class fitted. We implement such a decorator class in `AntiThetic.h` and `AntiThetic.cpp`.

Listing 6.5 (`AntiThetic.h`)

```
#ifndef ANTITHETIC_H
#define ANTITHETIC_H

#include <Random2.h>
#include <wrapper.h>

class AntiThetic : public RandomBase
{
```

```
public:
    AntiThetic(const Wrapper<RandomBase>& innerGenerator );

    virtual RandomBase* clone() const;

    virtual void GetUniforms(MJArray& variates);

    virtual void Skip(unsigned long numberOfPaths);

    virtual void SetSeed(unsigned long Seed);

    virtual void ResetDimensionality(unsigned long
                                        NewDimensionality);

    virtual void Reset();
private:
    Wrapper<RandomBase> InnerGenerator;

    bool OddEven;

    MJArray NextVariates;
};
#endif
```

The decorator class is quite simple. It has an array as a data member to store the last vector drawn, and a boolean to indicate whether the next draw should be drawn from the inner generator, or be the anti-thetic of the last draw. A copy of the generator we are using is stored using the `Wrapper` template class and cloning, as usual. Note that we are actually taking a copy of the generator here so that the sequence of draws from the original generator will not be affected by drawing from the anti-thetic generator.

Listing 6.6 (AntiThetic.cpp)

```
#include <AntiThetic.h>

AntiThetic::AntiThetic(const Wrapper<RandomBase>&
                                    innerGenerator )
        : RandomBase(*innerGenerator),
            InnerGenerator(innerGenerator)
{
```

```
    InnerGenerator->Reset();
    OddEven =true;
    NextVariates.resize(GetDimensionality());
}

RandomBase* AntiThetic::clone() const
{
    return new AntiThetic(*this);
}

void AntiThetic::GetUniforms(MJArray& variates)
{
    if (OddEven)
    {
        InnerGenerator->GetUniforms(variates);

        for (unsigned long i =0; i < GetDimensionality(); i++)
            NextVariates[i] = 1.0-variates[i];

        OddEven = false;
    }
    else
    {
        variates = NextVariates;

        OddEven = true;
    }
}

void AntiThetic::SetSeed(unsigned long Seed)
{
    InnerGenerator->SetSeed(Seed);
    OddEven = true;
}

void AntiThetic::Skip(unsigned long numberOfPaths)
{
    if (numberOfPaths ==0)
        return;

    if (OddEven)
```

```
    {
        OddEven = false;
        numberOfPaths--;
    }


    InnerGenerator->Skip(numberOfPaths  / 2);

    if (numberOfPaths % 2)
    {
        MJArray tmp(GetDimensionality());

        GetUniforms(tmp);

    }
}


void AntiThetic::ResetDimensionality(unsigned long
                                           NewDimensionality)
{
    RandomBase::ResetDimensionality(NewDimensionality);

    NextVariates.resize(NewDimensionality);

    InnerGenerator->ResetDimensionality(NewDimensionality);
}

void AntiThetic::Reset()
{
    InnerGenerator->Reset();
    OddEven =true;
}
```

The implementation of the class is quite straightforward. Most of the methods consist of simply forwarding the request to the inner class, together with book-keeping for odd and even paths. The main `GetUniforms` method, gets uniforms from the inner generator for the odd draws, stores the results, X_j, and returns $(1 - X_1, \ldots, 1 - X_n)$ for the even draws. Note that

$$N^{-1}(1 - x) = -N^{-1}(x), \tag{6.1}$$

so this will yield the negative of the Gaussian variates if the `GetGaussians` method is used, as we wanted.

Note the syntax for initialization in the constructor. We have RandomBase
(*innerGenerator). As innerGenerator is a wrapped pointer, * gives us the
value of the inner object which is a member of some inherited class. However,
we can always treat any inherited class object as a base class object so the call
to RandomBase invokes the base class copy constructor, copying the base class
data in innerGenerator, and thus ensuring that the new object has the correct
dimensionality stored.

6.6 Using the random number generator class

Now that we have a random number generator class, we need to adapt our Monte
Carlo code to work with it. We give an adapted vanilla option pricer in
SimpleMC8.h and SimpleMC8.cpp. The header file declares the new func-
tion.

Listing 6.7 (SimpleMC8.h)

```
#ifndef SIMPLEMC8_H
#define SIMPLEMC8_H

#include <Vanilla3.h>
#include <Parameters.h>
#include <Random2.h>
#include <MCStatistics.h>

void SimpleMonteCarlo6(const VanillaOption& TheOption,
                       double Spot,
                       const Parameters& Vol,
                       const Parameters& r,
                       unsigned long NumberOfPaths,
                       StatisticsMC& gatherer,
                       RandomBase& generator);
#endif
```

We have chosen to take the random number generator in as a non-const reference.
It cannot be a const reference as the act of drawing a random number changes the
generator and is therefore implemented by a non-const method. The effect of this
is that any random numbers drawn inside the function will not be produced outside
the function, but instead the generator will continue where the function left off.
If we wanted the generator to be totally unaffected by what happened inside the
function, we would change the function to take in the object by value instead. Or
alternatively, we could copy the object and pass in the copy to the function, which

would have the same net effect. As usual, we use a reference to the base class in
order to allow the caller to decide how to implement the generator.

The implementation is as follows:

Listing 6.8 (SimpleMC8.cpp)

```
#include<SimpleMC8.h>
#include <cmath>
#include <Arrays.h>

// the basic math functions should be in
// namespace std but aren't in VCPP6
#if !defined(_MSC_VER)
using namespace std;
#endif

void SimpleMonteCarlo6(const VanillaOption& TheOption,
                       double Spot,
                       const Parameters& Vol,
                       const Parameters& r,
                       unsigned long NumberOfPaths,
                       StatisticsMC& gatherer,
                       RandomBase& generator)
{
    generator.ResetDimensionality(1);

    double Expiry = TheOption.GetExpiry();
    double variance = Vol.IntegralSquare(0,Expiry);
    double rootVariance = sqrt(variance);
    double itoCorrection = -0.5*variance;
    double movedSpot = Spot*exp(r.Integral(0,Expiry)
                      + itoCorrection);

    double thisSpot;
    double discounting = exp(-r.Integral(0,Expiry));

    MJArray VariateArray(1);

    for (unsigned long i=0; i < NumberOfPaths; i++)
    {
```

```
    generator.GetGaussians(VariateArray);
    thisSpot = movedSpot*exp(rootVariance*VariateArray[0]);
    double thisPayOff = TheOption.OptionPayOff(thisSpot);
    gatherer.DumpOneResult(thisPayOff*discounting);
}


    return;
}
```

We only comment on the new aspects of the routine. We first reset the generator's dimensionality to 1 as pricing a vanilla option is a one-dimensional integral – we just need the location of the final value of spot.

We set up the array in which to store the variate before we set up the main loop, once and for all. This avoids any difficulties with speed in the allocation of dynamically sized arrays. The GetGaussians method of the generator is used to write the variates (in this case just one variate, of course) into the array. This variate is then used as before to compute the final value of spot.

We give an example of using this routine with anti-thetic sampling in Random-Main3.cpp.

Listing 6.9 (RandomMain3.cpp)

```
/*
    uses source files
    AntiThetic.cpp
    Arrays.cpp,
    ConvergenceTable.cpp,
    MCStatistics.cpp
    Normals.cpp
    Parameters.cpp,
    ParkMiller.cpp
    PayOff3.cpp,
    PayOffBridge.cpp,
    Random2.cpp,
    SimpleMC8.cpp
    Vanilla3.cpp,
*/
#include<SimpleMC8.h>
#include<ParkMiller.h>
#include<iostream>
```

```cpp
using namespace std;
#include<Vanilla3.h>
#include<MCStatistics.h>
#include<ConvergenceTable.h>
#include<AntiThetic.h>

int main()
{

    double Expiry;
    double Strike;
    double Spot;
    double Vol;
    double r;
    unsigned long NumberOfPaths;

    cout << "\nEnter expiry\n";
    cin >> Expiry;

    cout << "\nStrike\n";
    cin >> Strike;

    cout << "\nEnter spot\n";
    cin >> Spot;

    cout << "\nEnter vol\n";
    cin >> Vol;

    cout << "\nr\n";
    cin >> r;

    cout << "\nNumber of paths\n";
    cin >> NumberOfPaths;

    PayOffCall thePayOff(Strike);

    VanillaOption theOption(thePayOff, Expiry);

    ParametersConstant VolParam(Vol);
    ParametersConstant rParam(r);
```

```
StatisticsMean gatherer;
ConvergenceTable gathererTwo(gatherer);

RandomParkMiller generator(1);

AntiThetic GenTwo(generator);

SimpleMonteCarlo6(theOption,
                            Spot,
                            VolParam,
                            rParam,
                            NumberOfPaths,
                            gathererTwo,
                            GenTwo);

vector<vector<double> > results =
    gathererTwo.GetResultsSoFar();

cout <<"\nFor the call price the results are \n";
for (unsigned long i=0; i < results.size(); i++)
{
    for (unsigned long j=0; j < results[i].size(); j++)
        cout << results[i][j] << " ";

    cout << "\n";
}
double tmp;
cin >> tmp;

return 0;
}
```

We create a Park–Miller random number generator object and then wrap it with an anti-thetic decorator. This decorated object is then passed into the new Monte Carlo routine. As usual, the routine is not aware of the fact that the passed-in object has been decorated but simply uses it in the same way as any other random number generator.

The big difference between our new program and the old ones is that the results are now compiler-independent. The numbers returned are now precisely the same under Borland 5.5, Visual C++ 6.0 and MingW 2.95, since we have removed the dependency on the inbuilt `rand()` function which previously made our results compiler-dependent. This gives us an extra robustness test; if our results are not now compiler-independent we should be worried and find out why!

6.7 Key points

In this chapter, we developed a random number generator class and saw how anti-thetic sampling could be implemented via decoration.

- `rand` is implementation-dependent.
- Results from `rand()` are not easily reproducible.
- We have to be sure that a random generator is capable of the dimensionality necessary for a simulation.
- Using a random number class allows us to use decoration.
- The inverse cumulative normal function is the most robust way to turn uniform variates from the open interval, (0, 1), into Gaussian variates.
- Using a random number class makes it easier to plug in low-discrepancy numbers.
- Anti-thetic sampling can be implemented via decoration.

6.8 Exercises

Exercise 6.1 For various cases compare convergence of Monte Carlo simulations with and without anti-thetic sampling.

Exercise 6.2 Obtain another random number generator and fit it into the class hierachy given here. (See [26] or `www.boost.org` for other generators.)

Exercise 6.3 Code up a low-discrepancy number generator and integrate it into the classes here. (See [26] or [10].)

7

An exotics engine and the template pattern

7.1 Introduction

We have now developed quite a few sets of components: random number generators, parameters classes, pay-off classes, statistics gatherers and a wrapper template. Having developed all these components with the objective of reusability, we examine in this chapter how to put them together to price path-dependent exotic options. Our objective is to develop a flexible Monte Carlo pricer for exotic options which pay off at some future date according to the value of spot on a finite number of dates. We will work within a deterministic interest rate world, and assume the Black–Scholes model of stock price evolution.

We assume that our derivative is discrete, i.e. that it depends upon the value of spot on a discrete set of times. Thus our derivative is associated to a set of times, $t_1, t_2, \ldots, t_n$, and pays at some time T a function $f(S_{t_1}, \ldots, S_{t_n})$ of the value of spot at those times. For example, a one-year Asian call option struck at K with monthly resets would pay

$$\left(\frac{1}{12} \sum_{j=1}^{12} S_{t_j} - K \right)_+ ,$$

where $t_j = j/12$, at time 1.

More generally, the derivative could possibly pay cash-flows at more than one time. For example, a discrete barrier knock-out option could pay an ordinary vanilla pay-off at the time of expiry, and a rebate at the time of knock-out otherwise.

We do not consider American or Bermudan options here as the techniques involved are quite different. Note, however, that once an exercise strategy has been chosen the option is really just a path-dependent derivative and so the option can be evaluated by these techniques for any given fixed exercise strategy.

7.2 Identifying components

Before designing our engine let's identify what it will have to do. For each path, we generate a discounted pay-off which is then averaged over all paths to obtain a price. To generate a pay-off for a path, we have to know the path of stock prices at the relevant times, plug this path into the pay-off function of the option to obtain the cash-flows, and then discount these cash-flows back to the start to obtain the price for that path.

We can therefore identify four distinct actions:

(i) the generation of the stock price path;
(ii) the generation of cash-flows given a stock price path;
(iii) the discounting and summing of cash-flows for a given path;
(iv) the averaging of the prices over all the paths.

We already have a suitable component for the last of these actions: the statistics gatherer. We will just plug in the class we have already written at the appropriate point. For the second action, we are purely using the definition of the derivative to determine what its pay-off is, given a set of stock prices. An obvious component for our model is therefore a path-dependent exotic option class which will encapsulate the information which would be written in the *term-sheet* for the option.

Note that by defining the concept we model by the term-sheet, we make it clear that the class will not involve interest rates nor knowledge of volatility nor any aspect of the stock price process. A consequence of this is that the option class can only ever say what cash-flows occur and when: it cannot say anything about their discounted values because that would require knowledge of interest rates. Note the general point here that, by defining the concept in real-world terms, it becomes very easy to decide what should and should not be contained in the class. Another consequence is that the component is more easily reusable: if we decide to do jump-diffusion pricing or stochastic interest rates, this class will be reusable without modification, and that would not be the case if we had included information about interest rates or volatility.

There is more than one way to handle the two remaining tasks: path generation and cash-flow accounting. The latter will be the same for any deterministic interest-rate model and it is therefore natural to include it as part of our main engine class. We can require path generation to be an input to our main class, and therefore define it in terms of its own class hierarchy, or via a virtual method of the base class. A third option, which we do not explore, would be to make it a template parameter, which would avoid the small overhead of a virtual function call.

The option we will pursue here is to make path generation a virtual method of the base class. This is an example of the *template* design pattern, which should not be confused with templatized code. The idea here is that the base class sets up a structure with methods that control everything – in this case it would be methods to run the simulation and account for each path – though the main work of actually generating the path is not defined in the base class, but is instead called via a pure virtual function. This pure virtual function must therefore be defined in an inherited class. Thus as in templatized code, the code is set up more to define a structure than to do the real work which is coded elsewhere.

7.3 Communication between the components

Having identified what our components will be, we still need to assess what information has to be passed between them, and we need to decide how to carry out the communication.

The product will take in a vector of spot values for its relevant times and spit out the cash-flows generated. A couple of immediate consequences of this are that there has to be a mechanism for the product to tell the path generator for which times it needs the value of spot, and that we need to decide how to define a cash-flow object.

To deal with the first of these, we include a method `GetLookAtTimes`; this passes back an array of times that are relevant to the pay-off function of the product. For the definition of cash-flow objects, there are a couple of options. The first obvious approach is to simply make a cash-flow a pair of `doubles` which define the amount and the time of the cash-flow. This approach has the disadvantage that if we have a complicated term structure of interest rates, the action of computing the discount factor for the cash-flow time may be slow, and with the product being allowed to pass back an arbitrary time, this discounting will have to be done on every path. Whilst one could cache already-computed times, there is then still the problem that searching the cache for the already-computed times will take time.

In practice, many products can only pay off at one time. This means that it would be better to pre-compute the discount factor for that time. However, we would still need to know in advance what that time is. We therefore require our product to have another method, `PossibleCashFlowTimes`, which returns an array defining the possible times. As the engine will know all the possible times in advance we can return a cash-flow as a pair: an index and an amount. The index is now an `unsigned long` instead of a `double`. The engine will now precompute all the discount factors and then simply use the index to look up an array to get the discounting for any given cash-flow.

We still have to decide the syntax for the main method `CashFlows`. The method takes in an array defining spot values, and returns cash-flows. As we allow the possibility of more than one cash-flow, we must use a container to pass them back. We use the STL vector class. Whilst it would be tempting to make the return type of the class `vector<CashFlow>`, this would have timing disadvantages. We would have to create a new vector every time the method was called, and this could be time consuming because any dynamically sized container class must involve memory allocation. We therefore take an argument of type `vector<CashFlow>&` into which we write the cash-flows.

We still have the issue that the `vector` will need to be of the correct size. One solution is for the method to resize it as necessary but this could be time consuming. First, resizing can involve memory allocation though this is not a huge issue since the memory allocated for an STL `vector` never shrinks so if the same `vector` is used every time it will rapidly grab enough memory and then will need no more. Second, some implementations of the STL explicitly destroy all the objects in the `vector` during a resize, which means that every resize involves looping, and is therefore unnecessarily slow even when no memory allocation is necessary.

The solution we adopt is to tell the outside engine how big the vector has to be, and then each time the method is called, to return an `unsigned long` saying how many cash-flows have been generated. Thus we have two pure virtual methods:

```
virtual unsigned long MaxNumberOfCashFlows() const=0;
virtual unsigned long CashFlows(const MJArray& SpotValues,
                                std::vector<CashFlow>&
                                GeneratedFlows) const=0;
```

So in summary our objects will communicate as follows:

   (i) The path generator asks the product what times it needs spot for, and it passes back an array.

  (ii) The accounting part of the engine asks the product what cash-flow times are possible, and it passes back an array. The engine then computes all the possible discount factors.

 (iii) The accounting part of the engine asks the product what is the maximum number of cash flows it can generate, and sets up a vector of that size.

 (iv) For each path, the engine gets an array of spot values from the path generator.

  (v) The array of spot values is passed into the product, which passes back the number of cash-flows, and puts their values into the vector.

 (vi) The cash-flows are discounted appropriately and summed, and the total value is passed into the statistics gatherer.

(vii) After all the looping is done, the final results are obtained from the statistics gatherer.

7.4 The base classes

Having discussed in previous sections what classes will be needed and how they should communicate, in this section we give the implementations of the base classes.

In `PathDependent.h`, we define the `CashFlow` and the `PathDependent` classes which give our path-dependent exotic option.

Listing 7.1 (PathDependent.h)

```cpp
#ifndef PATH_DEPENDENT_H
#define PATH_DEPENDENT_H
#include <Arrays.h>
#include <vector>

class CashFlow
{
public:
    double Amount;
    unsigned long TimeIndex;

    CashFlow(unsigned long TimeIndex_=0UL, double Amount_=0.0)
                : TimeIndex(TimeIndex_),
                  Amount(Amount_){};
};

class PathDependent
{
public:
    PathDependent(const MJArray& LookAtTimes);

    const MJArray& GetLookAtTimes() const;

    virtual unsigned long MaxNumberOfCashFlows() const=0;
    virtual MJArray PossibleCashFlowTimes() const=0;
    virtual unsigned long CashFlows(const MJArray& SpotValues,
                                    std::vector<CashFlow>&
                                      GeneratedFlows) const=0;
    virtual PathDependent* clone() const=0;

    virtual ~PathDependent(){}
private:
    MJArray LookAtTimes;
};
#endif
```

The CashFlow class is really just a struct as it has public data members. Note that we ensure that it has a default constructor by giving the constructor default arguments, this is necessary in order to use it with STL containers which need a default constructor for certain operations such as creating a vector of arbitrary size.

The base class for `PathDependent` really does not do much except define the interface. We have made the base class store the `LookAtTimes` as every possible product will need these times, and provided the method `GetLookAtTimes` to obtain them. As usual, we include a `clone` method for virtual copy construction, and a virtual destructor to make sure that there are no memory leaks arising from destroying base class objects instead of inherited ones.

The source code is suitably short:

Listing 7.2 (`PathDependent.cpp`)

```
#include <PathDependent.h>

PathDependent::PathDependent(const MJArray& LookAtTimes_)
                        :   LookAtTimes(LookAtTimes_)
{}

const MJArray& PathDependent::GetLookAtTimes() const
{
    return LookAtTimes;
}
```

There is a bit more to the base class for the engine as it will actually handle the accounting for the cash-flows.

Listing 7.3 (`ExoticEngine.h`)

```
#ifndef EXOTIC_ENGINE_H
#define EXOTIC_ENGINE_H
#include <wrapper.h>
#include <Parameters.h>
#include <PathDependent.h>
#include <MCStatistics.h>
#include <vector>
class ExoticEngine
{
public:
    ExoticEngine(const Wrapper<PathDependent>&
                    The Product_, const Parameters& r_);

    virtual void GetOnePath(MJArray& SpotValues)=0;
```

```
    void DoSimulation(StatisticsMC& TheGatherer,
                      unsigned long NumberOfPaths);
    virtual ~ExoticEngine(){}
    double DoOnePath(const MJArray& SpotValues) const;

private:
    Wrapper<PathDependent> TheProduct;
    Parameters r;
    MJArray Discounts;
    mutable std::vector<CashFlow> TheseCashFlows;
};
#endif
```

The engine has four data members. The product is stored using the Wrapper template as we do not know its type. The interest rates are stored using the Parameters class which will allow us variable ones if we so desire. We also delegate computation of integrals to the Parameters class, and not have to worry about them here.

We have an array Discounts, which will be used to store the discount factors in order for the possible cash-flow times. Finally we have a mutable data member TheseCashFlows. This means that it can change value inside const member functions. The idea is that the variable is not really a data member, but instead it is a workspace variable: this it is faster to declare once and for all in the class definition. Remember that creating and destroying containers can be time-consuming so we design the class so that the vector is created once and for all at the beginning.

Note that we split our main method; it has two auxiliary methods, DoOnePath and GetOnePath. The second of these is pure virtual and therefore will be defined in an inherited class which will involve a choice of stochastic process and model. Note that this method is not constant as we will want a different set of spot values every time, and so it will necessarily change something about the state of the object. The other of the methods does everything necessary for one path given the spot values. This is const as turning spot values into prices is a purely functional action with no underlying changes. Both these methods pass arrays by reference in order to avoid any memory allocation. Note the implicit assumption that the array passed into GetOnePath is of the correct size.

The source code for implementing the base class is fairly simple and straightforward as all the hard work has been hived off into auxiliary classes.

Listing 7.4 (ExoticEngine.cpp)

```
#include <ExoticEngine.h>
#include <cmath>
```

```cpp
ExoticEngine::ExoticEngine(const Wrapper<PathDependent>&
                                 TheProduct_,
                                 const Parameters& r_)
              :
              TheProduct(TheProduct_),
              r(r_),
              Discounts(TheProduct_->PossibleCashFlowTimes())
{
    for (unsigned long i=0; i < Discounts.size(); i++)
        Discounts[i] = exp(-r.Integral(0.0, Discounts[i]));

    TheseCashFlows.resize(TheProduct->MaxNumberOfCashFlows());
}

void ExoticEngine::DoSimulation(StatisticsMC& TheGatherer,
                                 unsigned long NumberOfPaths)
{
    MJArray SpotValues(TheProduct->GetLookAtTimes().size());

    TheseCashFlows.resize(TheProduct->MaxNumberOfCashFlows());
    double thisValue;

    for (unsigned long i =0; i < NumberOfPaths; ++i)
    {
        GetOnePath(SpotValues);
        thisValue = DoOnePath(SpotValues);
        TheGatherer.DumpOneResult(thisValue);
    }

    return;
}

double ExoticEngine::DoOnePath(const MJArray&
                                 SpotValues) const
{
    unsigned long NumberFlows =
            TheProduct->CashFlows(SpotValues,
                                     TheseCashFlows);
    double Value=0.0;
```

```
for (unsigned i =0; i < NumberFlows; ++i)
    Value += TheseCashFlows[i].Amount *
            Discounts[TheseCashFlows[i].TimeIndex];
```

```
    return Value;
}
```

The constructor stores the inputs, computes the discount factors necessary, and makes sure the cash-flows vector is of the correct size. The `DoSimulation` method loops through all the paths, calling `GetOnePath` to get the array of spot value and then passes them into `DoOnePath` to get the value for that set of spot values. This value is then dumped into the statistics gatherer.

DoOnePath is only slightly more complicated. The array of spot values is passed into the product to get the cash-flows. These cash-flows are then looped over and discounted appropriately. The discounting is simplified by using the precomputed discount factors.

We have now set up the structure for pricing path-dependent exotic derivatives but we still have to actually define the classes which will do the path generation and define the products.

7.5 A Black–Scholes path generation engine

The Black–Scholes engine will produce paths from the risk-neutral Black–Scholes process. The paths will be an array of spot values at the times specified by the product. We allow the possibility of variable interest rates and dividend rates, as well as variable but deterministic volatility. The stock price therefore follows the process

$$dS_t = (r(t) - d(t))S_t dt + \sigma(t)S_t dW_t, \tag{7.1}$$

with S_0 given. To simulate this process at times $t_0, t_1, \ldots, t_{n-1}$, we need n independent $N(0, 1)$ variates W_j and we set

$$\log S_{t_0} = \log S_0 + \int_0^{t_0} \left(r(s) - d(s) - \frac{1}{2}\sigma(s)^2 \right) ds + \sqrt{\int_0^{t_0} \sigma(s)^2 ds}\, W_0, \tag{7.2}$$

and put

$$\log S_{t_j} = \log S_{t_{j-1}} + \int_{t_{j-1}}^{t_j} \left(r(s) - d(s) - \frac{1}{2}\sigma(s)^2 \right) ds + \sqrt{\int_{t_{j-1}}^{t_j} \sigma(s)^2 ds}\, W_j. \tag{7.3}$$

We implement this procedure in `ExoticBSEngine.h` and `ExoticBS-Engine.cpp`.

Listing 7.5 (ExoticBSEngine.h)

```
#ifndef EXOTIC_BS_ENGINE_H
#define EXOTIC_BS_ENGINE_H
#include <ExoticEngine.h>
#include <Random2.h>

class ExoticBSEngine : public ExoticEngine
{
public:
    ExoticBSEngine(const Wrapper<PathDependent>& TheProduct_,
                   const Parameters& R_,
                   const Parameters& D_,
                   const Parameters& Vol_,
                   const Wrapper<RandomBase>& TheGenerator_,
                   double Spot_);

    virtual void GetOnePath(MJArray& SpotValues);
    virtual ~ExoticBSEngine(){}

private:
    Wrapper<RandomBase> TheGenerator;
    MJArray Drifts;
    MJArray StandardDeviations;
    double LogSpot;
    unsigned long NumberOfTimes;
    MJArray Variates;
};
#endif
```

Listing 7.6 (ExoticBSEngine.cpp)

```
#include <ExoticBSEngine.h>
#include <cmath>

void ExoticBSEngine::GetOnePath(MJArray& SpotValues)
{

    TheGenerator->GetGaussians(Variates);
```

```
    double CurrentLogSpot = LogSpot;

    for (unsigned long j=0; j < NumberOfTimes; j++)
    {
        CurrentLogSpot += Drifts[j];
        CurrentLogSpot += StandardDeviations[j]*Variates[j];
        SpotValues[j] = exp(CurrentLogSpot);
    }

    return;
}


ExoticBSEngine::ExoticBSEngine(const Wrapper<PathDependent>&
                                                    TheProduct_,
                        const Parameters& R_,
                        const Parameters& D_,
                        const Parameters& Vol_,
                        const Wrapper<RandomBase>&
                                TheGenerator_,
                        double Spot_)
                        :
                        ExoticEngine(TheProduct_,R_),
                        TheGenerator(TheGenerator_)
{
    MJArray Times(TheProduct_->GetLookAtTimes());
    NumberOfTimes = Times.size();

    TheGenerator->ResetDimensionality(NumberOfTimes);
    Drifts.resize(NumberOfTimes);
    StandardDeviations.resize(NumberOfTimes);

    double Variance = Vol_.IntegralSquare(0,Times[0]);

    Drifts[0] = R_.Integral(0.0,Times[0])
            - D_.Integral(0.0,Times[0]) - 0.5 * Variance;
    StandardDeviations[0] = sqrt(Variance);

    for (unsigned long j=1; j < NumberOfTimes; ++j)
    {
```

```
        double thisVariance =
            Vol_.IntegralSquare(Times[j-1],Times[j]);
        Drifts[j] = R_.Integral(Times[j-1],Times[j])
                    - D_.Integral(Times[j-1],Times[j])
                    - 0.5 * thisVariance;
        StandardDeviations[j] = sqrt(thisVariance);
    }

    LogSpot = log(Spot_);
    Variates.resize(NumberOfTimes);
}
```

The integrals and square-roots are the same for every path and so can be precomputed. The constructor therefore gets the times from the product, and uses them to compute the integrals of the drifts and the standard deviations which are stored as data members. Note that the class does not bother to store the times as it is only the constructor which needs to know what they are. In any case, the product is passed up to the base class and it could be retrieved from there if it were necessary.

The generation will of course require a random number generator and we pass in a wrapped `RandomBase` object to allow us to plug in any one we want without having to do any explicit memory handling. We have a data member `Variates` so that the array can be defined once and for all at the beginning: once again this is with the objective of avoiding unnecessary creation and deletion of objects. We store the log of the initial value of spot as this is the most convenient for carrying out the path generation.

As we have done a lot of precomputation in the constructor, the routine to actually generate a path is fairly simple. We simply get the variates from the generator and loop through the times. For each time, we add the integrated drift to the log, and then add the product of the random number and the standard deviation. To minimize the number of calls to log and exp, we keep track of the log of the spot at all times, and convert into spot values as necessary. We thus have `NumberOfTimes` calls to exp each path and no calls to log. As we will have to exponentiate to change our Gaussian into a log-normal variate at some point, this appears to be optimal for this design. If we were really worried that too much time was being spent on computing exponentials, one solution would be to change the design and pass the log of the values of spot back, and then pass these log values into the product. The product would then have the obligation to exponentiate them if necessary. For certain products such as a geometric Asian option this might well be faster as it would only involve one exponentiation instead of many. The main downside would be that for certain processes, such as a normal process or displaced diffusion, one might end up having to take unnecessary logs.

7.6 An arithmetic Asian option

Before we can run our engine, we need one last thing, namely a concrete product to put in it. One simple example is an arithmetic Asian option. Rather than define a different class for each sort of pay-off, we use the already developed `PayOff` class as a data member.

The header file for the class is quite simple:

Listing 7.7 (PathDependentAsian.h)

```
#ifndef PATH_DEPENDENT_ASIAN_H
#define PATH_DEPENDENT_ASIAN_H

#include <PathDependent.h>
#include <PayOffBridge.h>

class PathDependentAsian : public PathDependent
{
public:
    PathDependentAsian(const MJArray& LookAtTimes_,
                       double DeliveryTime_,
                       const PayOffBridge& ThePayOff_);

    virtual unsigned long MaxNumberOfCashFlows() const;
    virtual MJArray PossibleCashFlowTimes() const;
    virtual unsigned long CashFlows(const MJArray& SpotValues,
                std::vector<CashFlow>& GeneratedFlows) const;
    virtual ~PathDependentAsian(){}
    virtual PathDependent* clone() const;
private:
    double DeliveryTime;
    PayOffBridge ThePayOff;
    unsigned long NumberOfTimes;
};
#endif
```

The methods defined are just the ones required by the base class. We pass in the averaging times as an array and we provide a separate delivery time to allow for the possibility that the pay-off occurs at some time after the last averaging date. Note that the use of PayOffBridge class means that the memory handling is handled internally, and this class does not need to worry about assignment, copying and destruction.

The source file is fairly simple too.

Listing 7.8 (PathDependentAsian.cpp)

```
#include <PathDependentAsian.h>


PathDependentAsian::PathDependentAsian(const MJArray&
                                        LookAtTimes_,
                          double DeliveryTime_,
                          const PayOffBridge&ThePayOff_)
                          :
                          PathDependent(LookAtTimes_),
                          DeliveryTime(DeliveryTime_),
                          ThePayOff(ThePayOff_),
                          NumberOfTimes(LookAtTimes_.size())
{
}


unsigned long PathDependentAsian::MaxNumberOfCashFlows() const
{
    return 1UL;
}


MJArray PathDependentAsian::PossibleCashFlowTimes() const
{
    MJArray tmp(1UL);
    tmp[0] = DeliveryTime;
    return tmp;
}


unsigned long PathDependentAsian::CashFlows(const MJArray&
                                        SpotValues,
            std::vector<CashFlow>& GeneratedFlows) const
{
    double sum = SpotValues.sum();
    double mean = sum/NumberOfTimes;

    GeneratedFlows[0].TimeIndex = 0UL;
```

```
GeneratedFlows[0].Amount = ThePayOff(mean);

    return 1UL;
}


PathDependent* PathDependentAsian::clone() const
{
    return new PathDependentAsian(*this);
}
```

Note that our option only ever returns one cash-flow so the maximum number of cash-flows is 1. It only ever generates cash-flows at the delivery time so the `PossibleCashFlowTimes` method is straightforward too. The `CashFlows` method takes the spot values, sums them, divides by the number of them and calls `ThePayOff` to find out what the pay-off is. The answer is then written into the `GeneratedFlows` array and we are done.

7.7 Putting it all together

We now have everything we need to price an Asian option. We give an example of a simple interface program in `EquityFXMain.cpp`.

Listing 7.9 (`EquityFXMain.cpp`)

```
/*
    uses source files
    AntiThetic.cpp
    Arrays.cpp,
    ConvergenceTable.cpp,
    ExoticBSEngine.cpp
    ExoticEngine.cpp
    MCStatistics.cpp
    Normals.cpp
    Parameters.cpp,
    ParkMiller.cpp,
    PathDependent.cpp
    PathDependentAsian.cpp
    PayOff3.cpp,
    PayOffBridge.cpp,
    Random2.cpp,
*/
```

```cpp
#include<ParkMiller.h>
#include<iostream>
using namespace std;
#include<MCStatistics.h>
#include<ConvergenceTable.h>
#include<AntiThetic.h>
#include<PathDependentAsian.h>
#include<ExoticBSEngine.h>
int main()
{
    double Expiry;
    double Strike;
    double Spot;
    double Vol;
    double r;
    double d;
    unsigned long NumberOfPaths;
    unsigned NumberOfDates;

    cout << "\nEnter expiry\n";
    cin >> Expiry;

    cout << "\nStrike\n";
    cin >> Strike;

    cout << "\nEnter spot\n";
    cin >> Spot;

    cout << "\nEnter vol\n";
    cin >> Vol;

    cout << "\nr\n";
    cin >> r;

    cout << "\nd\n";
    cin >> d;

    cout << "Number of dates\n";
    cin >> NumberOfDates;
```

```
cout << "\nNumber of paths\n";
cin >> NumberOfPaths;


PayOffCall thePayOff(Strike);


MJArray times(NumberOfDates);


for (unsigned long i=0; i < NumberOfDates; i++)
    times[i] = (i+1.0)*Expiry/NumberOfDates;


ParametersConstant VolParam(Vol);
ParametersConstant rParam(r);
ParametersConstant dParam(d);


PathDependentAsian theOption(times, Expiry, thePayOff);


StatisticsMean gatherer;
ConvergenceTable gathererTwo(gatherer);


RandomParkMiller generator(NumberOfDates);


AntiThetic GenTwo(generator);


ExoticBSEngine theEngine(theOption, rParam, dParam,
                         VolParam, GenTwo, Spot);


theEngine.DoSimulation(gathererTwo, NumberOfPaths);


vector<vector<double> > results =
    gathererTwo.GetResultsSoFar();


cout <<"\nFor the Asian call price the results are \n";


{
for (unsigned long i=0; i < results.size(); i++)
   {
   for (unsigned long j=0; j < results[i].size(); j++)
        cout << results[i][j] << " ";

   cout << "\n";
```

```
        }}

        double tmp;
        cin >> tmp;

        return 0;
}
```

7.8 Key points

In this chapter, we saw how we can put the ideas developed in the previous chapters together to build a pricer for exotic options.

- An important part of the design process is identifying the necessary components and specifying how they talk to each other.
- The template pattern involves deferring the implementation of an important part of an algorithm to an inherited class.
- If an option class knows nothing that is not specified in the term-sheet then it is much easier to reuse.
- We can reuse the `PayOff` class to simplify the coding of our more complicated path-dependent derivatives.

7.9 Exercises

Exercise 7.1 Write a class to do geometric Asian options.

Exercise 7.2 Write a class to do discrete knock-out options that pay a rebate at the time of rebate.

Exercise 7.3 Rewrite the classes here so that they pass the logs of spot values around instead of the spot values. Show that the discrete barrier option and the geometric Asian need fewer exponentiations.

Exercise 7.4 Implement an engine for pricing when the spot price is normal instead of log-normal.

Exercise 7.5 Write a class that pays the difference in pay-offs of two arbitrary path-dependent derivatives.

8

Trees

8.1 Introduction

We have studied Monte Carlo code in detail: now we examine how we can apply similar techniques to pricing on trees. Before we can start designing the code, we need to fix the underlying mathematics. The point of view we adopt is that a tree is a method of approximating the risk-neutral expectation. In particular, we assume that we are pricing a derivative on a stock following geometric Brownian motion with a constant volatility σ. We let the continuously compounding interest rate be r and the continuous dividend rate be d. The dynamics for the stock in the risk-neutral measure are therefore

$$dS = (r - d)S dt + \sigma S dW_t. \tag{8.1}$$

The value of a European option with expiry T is then

$$e^{-rT} \mathbb{E}(C(S, T)), \tag{8.2}$$

where $C(S, T)$ is the final pay-off.

When we price on a binomial tree, we divide time into steps and across each step we assume that the underlying Brownian motion can only move a fixed amount up or down. The dynamics of the stock price under geometric Brownian motion are such that

$$S_t = S_0 e^{(r-d-\frac{1}{2}\sigma^2)t + \sigma W_t}. \tag{8.3}$$

We wish to discretize W_t. We have N steps to get from 0 to T. Each time step is therefore of length T/N. Across step l, we need to approximate

$$W_{(l+1)T/N} - W_{lT/N} = \sqrt{\frac{T}{N}} N(0, 1). \tag{8.4}$$

There is only one random variable taking precisely two values which has the same mean and variance as $N(0, 1)$, and this variable takes ± 1 with probability $1/2$. We therefore take a set of N independent random variables X_j with this distribution, and approximate $W_{lT/N}$ by

$$Y_l = \sqrt{\frac{T}{N}} \sum_{j=1}^{l} X_j. \tag{8.5}$$

The approximation for $S_{lT/N}$ is then

$$S_0 e^{(r-d-\frac{1}{2}\sigma^2)lT/N + \sigma Y_l}.$$

Note the crucial point here that since the value of S_t does not depend on the path of W_t but solely upon its value at time t, it is only the value of Y_l that matters not the value of each individual X_j. This is crucial because it means that our tree is recombining; it does not matter whether we go down then up, or up then down. This is not the case if we allow variable volatility, which is why we have assumed its constancy.

The nature of martingale pricing means that the value at a given time-step is equal to the discounted value at the next time-step, thus if we let $S_{l,k}$ be the value of the stock at time Tl/N if Y_l is k, we have that

$$
\begin{aligned}
C(S_{l,k}, Tl/N) &= e^{-rT/N} \mathbb{E}(S_{l+1}(Y_{l+1}|Y_l = k)), \\
&= \frac{1}{2} e^{-rT/N} \Big(C\big(S_{l,k} e^{(r-d-\frac{1}{2}\sigma^2)T/N + \sigma\sqrt{T/N}}, (l+1)T/N \big) \\
&\quad + C\big(S_{l,k} e^{(r-d-\frac{1}{2}\sigma^2)T/N - \sigma\sqrt{T/N}}, (l+1)T/N \big) \Big). \tag{8.6}
\end{aligned}
$$

Note that we are not doing true martingale pricing in the discrete world in that we are not adjusting probabilities to make sure the assets are discrete martingales; we are instead approximating the continuous martingales with discrete random variables which are almost, but not quite, martingales.

What does all this buy us? The value of $C(S_{N,k})$ is easy to compute for any value of k: we just plug the stock price into the final pay-off. Since we have a formula that expresses the value at step l into that at step $l+1$, we can now just backwards iterate through the tree in order to get the price at time zero.

However, the purpose of a tree is not to price a European option; there are lots of better ways of doing that, including analytical solutions and numerical integration. The reason trees were introduced was that they give an effective method for pricing American options. The analysis for an American option is similar except that the value at a point in the tree is the maximum of the exercise value at that point and the discounted expected value at the next time. This corresponds to the optimal strategy of exercise if and only if exercise gives more money than not exercising.

Our algorithm for pricing an American option is therefore as follows:

(i) Create an array of final spot values which are of the form

$$S_0 e^{(r-d-\frac{1}{2}\sigma^2)T+\sigma\sqrt{T/N}j}$$

where j ranges from $-N$ to N.

(ii) For each of these spot values evaluate the pay-off and store it.

(iii) At the previous time-slice compute the possible values of spot: these will be of the form

$$S_0 e^{(r-d-\frac{1}{2}\sigma^2)(N-1)T/N+\sigma\sqrt{T/N}j},$$

where j ranges from $-(N-1)$ to $N-1$.

(iv) For each of these values of spot, compute the pay-off and take the maximum with the discounted pay-off of the two possible values of spot at the next time.

(v) Repeat 3,4 until time zero is reached.

What else could we price on a tree? We could do a barrier option or an American option that could only be exercised within certain time periods. For a knock-out barrier option, the procedure would be the same as for the European, except that the value at a point in the tree would be zero if it lay behind the barrier. For an American option with limited early exercise the procedure would be the same again, except that we would only take the maximum at times at which early exercise was allowed. So in each case, when we change the option, all that alters is the rule for updating the value at a point in the tree.

Note that in our formulation, we have not used any no-arbitrage arguments. The reason is that we have implicitly assumed that the no-arbitrage arguments have been done in the continuous case before any discretization has been carried out. This means that the completeness property of a binomial tree, i.e. that it generates a *unique* no-arbitrage price, is not relevant. In particular, we could replace X_j by any random variable with mean 0 and variance 1. If we use X'_j which takes values $-\sqrt{2}, 0, \sqrt{2}$ with probabilities $0.25, 0.5$ and 0.25 respectively, then we obtain a trinomial tree on which we could carry out a very similar analysis and algorithm. We could in fact go much further and have as many values as we wanted as long as we took care to make sure that the tree still recombined.

8.2 The design

Having re-examined the mathematics and the algorithms, we are now in a position to think about the design. Here are some concepts that our discussion has thrown up:

- the discretization;
- the final pay-off of the option;

- the rule for deciding the value of an option at a point in the tree given spot and the discounted future value of the option.

The first of these concepts decides the shape of the tree, whereas the second and third are properties of the option. There is thus an obvious orthogonalization: we have a tree class which handles the discretization, and a second class which deals with the final pay-off and the rule at previous times. In fact, we have already developed a class, `PayOff`, to encapsulate vanilla option pay-offs and it is ideal for reuse here.

There are a number of ways we could approach the second class. We could inherit from `PayOffBridged` since we could view our class as adding structure to an existing class. Whilst this would work in code, I personally dislike it as an option being priced on a tree is not a type of pay-off, and so the inheritance is not expressing an *is a* relationship. Another approach might be simply to define a second class to do the calculation rule, and plug both the final pay-off and the calculation rule into the tree. Since for American options the final pay-off is generally relevant at all times, such an approach seems sub-optimal as it might require two copies of the pay-off.

Ultimately, the pay-off is an aspect of the option, and it therefore makes more sense to define it as data member of the class which can referenced via a final pay-off method. Thus we define options on trees via an abstract base class which has three defining methods:

- `FinalTime` indicates the expiry time of the option;
- `FinalPayOff` gives the final pay-off as a function of spot;
- `PreFinalValue` gives the value at a point in the tree as a function of spot, time and the discounted future value of the option.

Note that by defining the option class in this fashion, we have not allowed it to know anything about interest rates nor the process of the underlying. This means it can be used in any tree-like structure provided the structure is always in a position to let it know its own discounted future value. Note the difference here between the option classes we are defining here and those we defined for Monte Carlo: whilst we are trying to encapsulate similar concepts, the difference is in the information we are able to feed in at a given time. For Monte Carlo, the entire history of spot is easy but the discounted future value of an option is hard, whereas on a tree the discounted future value is easy but the history is hard. However, both classes have easy access to the pay-off which means we are able to share the pay-off class.

Our other concept is the tree itself. The tree really has two aspects: the placement of the nodes of the tree and the computing of the option value at each of the nodes. Whilst we could further orthogonalize and define separate classes for each of these, we write a single class to do the binomial tree which takes in an option as an argument. An important point to note is that as the placement of the nodes does not depend upon the option, we can save ourselves time if we want to price several options by placing the nodes once and then pricing all the options on the same tree.

Given this fact, we design our tree class in such a way that the tree is built once, and then any option can be valued on the tree via a separate method.

8.3 The TreeProduct class

As we have decided to model the tree and the product separately, we develop a class hierarchy for the products we can value on trees. As usual, we use an abstract base class to define an interface. We define the class, TreeProduct, in TreeProducts.h.

Listing 8.1 (TreeProducts.h)

```
#ifndef TREE_PRODUCTS_H
#define TREE_PRODUCTS_H

class TreeProduct
{
public:
    TreeProduct(double FinalTime_);
    virtual double FinalPayOff(double Spot) const=0;
    virtual double PreFinalValue(double Spot,
                                 double Time,
                                 double DiscountedFutureValue)
                                const=0;
    virtual ~TreeProduct(){}
    virtual TreeProduct* clone() const=0;
    double GetFinalTime() const;

private:
    double FinalTime;
};
#endif
```

The only data member for the base class is FinalTime, and we provide a GetFinalTime to allow its value to be read. Note that this places the constraint on our products that they actually have a time of expiry! Thus we are implicitly disallowing perpetual options, however this is a good thing as it is not clear how to go about valuing such an option using a tree. Ultimately, one would have to approximate using a product with a finite expiry and it is probably better to do so explicitly than implicitly.

We provide the usual `clone` method and a virtual destructor to allow virtual copying, and to ensure the absence of memory leaks after virtual copying. The remaining methods are pure virtual and specify the value of the product at expiry and at previous times as we discussed above.

As most of the base class is abstract the source file is short:

Listing 8.2 (`TreeProducts.cpp`)

```
#include <TreeProducts.h>


TreeProduct::TreeProduct(double FinalTime_)
: FinalTime(FinalTime_)
{
}


double TreeProduct::GetFinalTime() const
{
    return FinalTime;
}
```

We provide below two concrete implementations of tree products: the European option and the American option. As we specify the pay-off using the `PayOff-Bridged` class, we do not need to write separate classes for calls and puts. The header files are straightforward, the only changes from the base class being the addition of a data member to specify the pay-off and the fact that virtual methods are now concrete instead of abstract. Note that everything to do with the expiry time is taken care of by the base class; the constructors need only pass it on to the base class.

Listing 8.3 (`TreeAmerican.h`)

```
#ifndef TREE_AMERICAN_H
#define TREE_AMERICAN_H

#include <TreeProducts.h>
#include <PayOffBridge.h>

class TreeAmerican : public TreeProduct
{

public:
    TreeAmerican(double FinalTime,
                 const PayOffBridge& ThePayOff_);
```

```
    virtual TreeProduct* clone() const;
    virtual double FinalPayOff(double Spot) const;
    virtual double PreFinalValue(double Spot,
                                double Time,
                                double DiscountedFutureValue)
                                const;
    virtual ~TreeAmerican(){}

private:
    PayOffBridge ThePayOff;

};
#endif
```

The header for the European option is very similar:

Listing 8.4 (TreeEuropean.h)

```
#ifndef TREE_EUROPEAN_H
#define TREE_EUROPEAN_H

#include <TreeProducts.h>
#include <PayOffBridge.h>

class TreeEuropean : public TreeProduct
{

public:
    TreeEuropean(double FinalTime,
                 const PayOffBridge& ThePayOff_);

    virtual TreeProduct* clone() const;
    virtual double FinalPayOff(double Spot) const;
    virtual double PreFinalValue(double Spot,
                                 double Time,
                                 double DiscountedFutureValue)
                                 const;
    virtual ~TreeEuropean(){}

private:
    PayOffBridge ThePayOff;
};
#endif
```

The source files are also straightforward.

Listing 8.5 (`TreeAmerican.cpp`)

```
#include <TreeAmerican.h>
#include <minmax.h>


TreeAmerican::TreeAmerican(double FinalTime,
                          const PayOffBridge& ThePayOff_)
               : TreeProduct(FinalTime),
                 ThePayOff(ThePayOff_)
{
}


TreeProduct* TreeAmerican::clone() const
{
    return new TreeAmerican(*this);
}


double TreeAmerican::FinalPayOff(double Spot) const
{
    return ThePayOff(Spot);
}


double TreeAmerican::PreFinalValue(double Spot,
                     double ,
// Borland compiler doesnt like unused named variables
                     double DiscountedFutureValue) const
{
    return max(ThePayOff(Spot), DiscountedFutureValue);
}
```

and

Listing 8.6 (`TreeEuropean.cpp`)

```
#include <TreeEuropean.h>
#include <minmax.h>


TreeEuropean::TreeEuropean(double FinalTime,
                          const PayOffBridge& ThePayOff_)
               : TreeProduct(FinalTime),
                 ThePayOff(ThePayOff_)
```

```
{
}

double TreeEuropean::FinalPayOff(double Spot) const
{
    return ThePayOff(Spot);
}

double TreeEuropean::PreFinalValue(double,
            //Spot,  Borland compiler
                            double,
        //Time,   doesnt like unused named variables
            double DiscountedFutureValue) const
{
    return DiscountedFutureValue;
}

TreeProduct* TreeEuropean::clone() const
{
    return new TreeEuropean(*this);
}
```

The implementations of the methods are very simple – all the pay-offs are sub-contracted to the `PayOffBridged` class in any case. For the European option at an interior node, the rule for computing the `PreFinalValue` is very simple: just return the discounted future value that was input. For the American option, it is only slightly harder; we take the maximum with the intrinsic value.

Note the slight oddity that we do not name the unused variables `Spot` and `Time` in the `PreFinalValue` method. This is because some compilers issue a warning if a variable is named but not used, to ensure that variables are not left unused accidentally.

8.4 A tree class

We give a simple implementation of a binomial tree class in this section. We design the tree to work in three pieces. The constructor does little except store the parameters. The `BuildTree` method actually makes the tree. In particular, it computes the locations of all the nodes, and the discounts needed to compute the expectations backwards in the tree. As it is not intended that the `BuildTree` method should be called from outside the class, we make it `protected` which allows the possibility that an inherited class may wish to use it without allowing any other external access.

The method which does the actual pricing is `GetThePrice`. Note that it takes in a `TreeProduct` by reference. As the argument is an abstract base class this means that an object from an inherited class must be passed in. Note that since we not need to store the object passed in, we do not need virtual constructors, wrappers or bridges. This method builds the tree if necessary, checks that the product has the right expiry time and then prices it. Our design is such that we can price multiple products with the same expiry; we call the method multiple times and only build the tree once. As we wish to be able to do this, we store the entire tree. Note that this is not really necessary since for any given time slice, we only need the next time slice and so we could easily save a lot of memory by only ever having two arrays defined. However, unless one is doing an awful lot of steps, memory will not be an issue, and this approach has the added benefit that if one wishes to analyze certain aspects of the product, such as where the exercise boundary lies for an American option, it is better to have the entire tree. We present the class in `BinomialTree.h`.

Listing 8.7 (`BinomialTree.h`)

```
#pragma warning( disable : 4786 )

#include <TreeProducts.h>
#include <vector>
#include <Parameters.h>
#include <Arrays.h>

class SimpleBinomialTree
{
public:
    SimpleBinomialTree(double Spot_,
                       const Parameters& r_,
                       const Parameters& d_,
                       double Volatility_,
                       unsigned long Steps,
                       double Time);

    double GetThePrice(const TreeProduct& TheProduct);

protected:
    void BuildTree();

private:
```

```
        double Spot;
        Parameters r;
        Parameters d;
        double Volatility;
        unsigned long Steps;
        double Time;
        bool TreeBuilt;

        std::vector<std::vector<std::pair<double, double> > >
                TheTree;
        MJArray Discounts;
};
```

Note that we store the tree as a vector of vectors of pairs of doubles. This is why we have the #pragma at the start; without the pragma, we get a warning message telling us that the debug info is too long (under Visual C++ 6.0).

A pair is a simple template class in the STL which simply gives a class with two data members of the appropriate types. They are accessed as public members as first and second. Note that an alternative implementation would be to have two trees: one for the spot and another for option values. However, that would require twice as much work when resizing, and more importantly one would then have to be careful to ensure that spot and the associated option value were always attached to the same indices. With the use of a pair, we get this for free.

We have allowed general Parameters for r and d: this is because variable interest and dividend rates change little in the analysis or construction of the tree. If we can add extra functionality with little cost, why not do so? We do not, however, allow variable volatility as it greatly complicates node placement; we would lose the property that the spot price is a simple function of the underlying Brownian motion.

We use a bool to indicate whether the tree has been built yet. We store the number of steps as an unsigned long and this is fixed in the constructor. One could easily add an additional method to allow the number of steps to be changed; however one would gain little over just instantiating a new object with a different number of steps. We have a data member for the discount factors needed so that they need only be computed once in BuildTree.

We present the source code in BinomialTree.cpp.

Listing 8.8 (BinomialTree.cpp)

```
#include <BinomialTree.h>
#include <Arrays.h>
#include <cmath>
```

```
// the basic math functions should be in namespace std
// but aren't in VCPP6
#if !defined(_MSC_VER)
using namespace std;
#endif

SimpleBinomialTree::SimpleBinomialTree(double Spot_,
                                       const Parameters& r_,
                                       const Parameters& d_,
                                       double Volatility_,
                                       unsigned long Steps_,
                                       double Time_)
                                       : Spot(Spot_),
                                       r(r_),
                                       d(d_),
                                       Volatility(Volatility_),
                                       Steps(Steps_),
                                       Time(Time_),
                                       Discounts(Steps)
{
    TreeBuilt=false;
}

void SimpleBinomialTree::BuildTree()
{
  TreeBuilt = true;
  TheTree.resize(Steps+1);

  double InitialLogSpot = log(Spot);

  for (unsigned long i=0; i <=Steps; i++)
  {

    TheTree[i].resize(i+1);

    double thisTime = (i*Time)/Steps;

    double movedLogSpot =
          InitialLogSpot + r.Integral(0.0, thisTime)
                        - d.Integral(0.0, thisTime);
```

```
    movedLogSpot -=
            0.5*Volatility*Volatility*thisTime;

    double sd = Volatility*sqrt(Time/Steps);

    for (long j = -static_cast<long>(i), k=0;
                j <= static_cast<long>(i); j=j+2,k++)
            TheTree[i][k].first = exp(movedLogSpot+ j*sd);
    }

    for (unsigned long l=0; l <Steps; l++)
    {
        Discounts[l] =
        exp(- r.Integral(l*Time/Steps,(l+1)*Time/Steps));
    }
}

double SimpleBinomialTree::GetThePrice(const TreeProduct&
                                                    TheProduct)
{
    if (!TreeBuilt)
        BuildTree();

    if (TheProduct.GetFinalTime() != Time)
        throw("mismatched product in SimpleBinomialTree");

    for (long j = -static_cast<long>(Steps), k=0;
            j <=static_cast<long>( Steps); j=j+2,k++)
        TheTree[Steps][k].second =
            TheProduct.FinalPayOff(TheTree[Steps][k].first);

    for (unsigned long i=1; i <= Steps; i++)
    {
        unsigned long index = Steps-i;
        double ThisTime = index*Time/Steps;

        for (long j = -static_cast<long>(index), k=0;
        j <= static_cast<long>(index); j=j+2,k++)
        {
            double Spot = TheTree[index][k].first;
```

```
        double futureDiscountedValue = 0.5*Discounts[index]*
        (TheTree[index+1][k].second +
          TheTree[index+1][k+1].second);
        TheTree[index][k].second =
        TheProduct.PreFinalValue(Spot,ThisTime,
                                     futureDiscountedValue);
    }


    }
    return TheTree[0][0].second;
}
```

The code is fairly straightforward. The constructor initializes the basic class variables and does not do much else. The method `BuildTree` creates the tree. We resize the `vector` describing all the layers first. We then resize each layer so that it is of the correct size. Note that as the number of nodes in a layer grows with the number of steps, these inner vectors are all of different sizes. We then compute a basepoint for the layer in log-space which is just the zero Brownian motion point. We then loop through the nodes in each layer writing in the appropriate value of spot.

Note that as we are dealing with points in the tree corresponding to down moves we count using a `long`. This `long` has to be compared to the `unsigned long i` for the termination condition. We therefore have to be careful; if we simply compare these numbers a likely effect is that the routine will conclude that -1 is bigger than 1. Why? The compiler will implicitly convert the `long` into a large `unsigned long`. Clearly this is not the desired effect so we convert the `unsigned long` into a `long` before the comparison.

After setting up the tree, we also set up the array of `Discounts` which are product-independent, and therefore only need to be done once.

The main routine for actually doing the pricing, `GetThePrice`, is also straightforward. We have put in a test to make sure that the tree and product are compatible. This throws an error which is a simple string if they are not. Note that this means the program will terminate unless we have written a `catch` which is capable of catching the string.

We simply iterate backwards through the tree. First we compute the final layer using the `FinalPayOff` and write the values into the `second` element of each `pair` in the final layer. After this we simply iterate backwards, computing as we go. The final value of the product is then simply the value of the option at the single node in the first layer of the tree, and that is what we return.

8.5 Pricing on the tree

Having written all the classes we need to actually put them together to price something. We give a simple interface in `TreeMain.cpp`.

Listing 8.9 (`TreeMain.cpp`)

```
/*
requires
    Arrays.cpp
    BinomialTree.cpp
    BlackScholesFormulas.cpp
    Normals.cpp
    Parameters.cpp
    PayOff3.cpp
    PayOffBridge.cpp
    PayOffForward.cpp
    TreeAmerican.cpp
    TreeEuropean.cpp
    TreeProducts.cpp
*/
#include <BinomialTree.h>
#include <TreeAmerican.h>
#include <TreeEuropean.h>
#include <BlackScholesFormulas.h>
#include <PayOffForward.h>
#include <iostream>
using namespace std;
#include <cmath>
int main()
{

    double Expiry;
    double Strike;
    double Spot;
    double Vol;
    double r;
    double d;
    unsigned long Steps;
```

```
cout << "\nEnter expiry\n";
cin >> Expiry;

cout << "\nStrike\n";
cin >> Strike;

cout << "\nEnter spot\n";
cin >> Spot;

cout << "\nEnter vol\n";
cin >> Vol;

cout << "\nr\n";
cin >> r;

cout << "\nd\n";
cin >> d;

cout << "\nNumber of steps\n";
cin >> Steps;

PayOffCall thePayOff(Strike);

ParametersConstant rParam(r);
ParametersConstant dParam(d);

TreeEuropean europeanOption(Expiry,thePayOff);
TreeAmerican americanOption(Expiry,thePayOff);

SimpleBinomialTree theTree(Spot,rParam,dParam,Vol,Steps,
                           Expiry);
double euroPrice = theTree.GetThePrice(europeanOption);
double americanPrice = theTree.GetThePrice(americanOption);
cout << "euro price" << euroPrice << "amer price"
                  << americanPrice << "\n";

double BSPrice = BlackScholesCall(Spot,Strike,r,d,
                                  Vol,Expiry);
cout << "BS formula euro price" << BSPrice << "\n";
```

```
PayOffForward forwardPayOff(Strike);
TreeEuropean forward(Expiry,forwardPayOff);

double forwardPrice = theTree.GetThePrice(forward);
cout << "forward price by tree" << forwardPrice << "\n";

double actualForwardPrice =
    exp(-r*Expiry)*(Spot*exp((r-d)*Expiry)-Strike);
cout << "forward price" << actualForwardPrice << "\n";

Steps++; // now redo the trees with one more step
SimpleBinomialTree theNewTree(Spot,rParam,dParam,Vol,
                                  Steps,Expiry);

double euroNewPrice =
    theNewTree.GetThePrice(europeanOption);
double americanNewPrice =
    theNewTree.GetThePrice(americanOption);

cout << "euro new price" << euroNewPrice
    << "amer new price" << americanNewPrice << "\n";

double forwardNewPrice = theNewTree.GetThePrice(forward);

cout << "forward price by new tree" << forwardNewPrice
                                << "\n";

double averageEuro = 0.5*(euroPrice+euroNewPrice);
double averageAmer = 0.5*(americanPrice+americanNewPrice);
double averageForward = 0.5*(forwardPrice+forwardNewPrice);

cout << "euro av price" << averageEuro << "amer av price"
                            << averageAmer << "\n";
cout << "av forward" << averageForward << "\n";

double tmp;
cin >> tmp;

return 0;
}
```

To illustrate certain aspects of tree pricing, we price the European call, the American call and the forward. We then reprice them using one extra step. The reason we do this is that often pricing on trees gives rise to a zig-zag behaviour as nodes go above and below the money. The average of the price for two successive steps is therefore often a lot more accurate than a single price. We give the comparison Black–Scholes price for the European option as an assessment of accuracy. The code for the Black–Scholes price is in `BlackScholesFormulas.h` and `BlackScholesFormulas.cpp`; we discuss it in Appendix A.

A standard way of improving the accuracy of American option pricing is to use the European price as a control. As we know the true price of the European and we know the tree price also, we can assume that the American option has the same amount of error as the European and adjust its price accordingly. The principle here is the same as that for a control variate in Monte Carlo simulation.

We also give the price of a forward. Note that if we had required the discounted discretized process to be a martingale, then the forward would be priced absolutely correctly; however as it is only an approximation to a martingale rather than actually being one, the forward need not be precise.

As we have not defined a class for the forward's pay-off before, we define one in `PayOffForward.h` and `PayOffForward.cpp`.

Listing 8.10 (`PayOffForward.h`)

```
#ifndef PAY_OFF_FORWARD_H
#define PAY_OFF_FORWARD_H

#include <PayOff3.h>
class PayOffForward : public PayOff
{
public:
    PayOffForward(double Strike_);

    virtual double operator()(double Spot) const;
    virtual ~PayOffForward(){}
    virtual PayOff* clone() const;

private:
    double Strike;
};
#endif
```

Listing 8.11 (`PayOffForward.cpp`)

```
#include <PayOffForward.h>

double PayOffForward::operator () (double Spot) const
{
    return Spot-Strike;
}


PayOffForward::PayOffForward(double Strike_) : Strike(Strike_)
{
}


PayOff* PayOffForward::clone() const
{
    return new PayOffForward(*this);
}
```

The class is straightforward and the only difference from the class defined for the call is that we take spot minus strike, instead of the call pay-off.

8.6 Key points

In this chapter we have used the patterns developed earlier in the book to develop routines for pricing on trees.

- Tree pricing is based on the discretization of a Brownian motion.
- Trees are a natural way to price American options.
- On a tree, knowledge of discounted future values is natural but knowing about the past is not.
- We can re-use the pay-off class when defining products on trees.
- By having a separate class encapsulating the definition of a derivative on a tree, we can re-use the products for more general structures.
- European options can be used as controls for American options.

8.7 Exercises

We have developed a very simple tree and treated a couple of simple products. The approach here can easily be extended to many more cases. We suggest a few possibilities for the reader to try.

Exercise 8.1 Find a class that does barrier options in the same `TreeProduct` class hierarchy. Try it out. How stable is the price? How might you improve the stability?

Exercise 8.2 Develop a binomial tree for which the memory requirements grow linearly with the number of steps. How do the memory requirements grow for the class here?

Exercise 8.3 Write a trinomial tree class.

Exercise 8.4 Modify the code so that it will work under variable volatility. The key is to ensure that the integral of the square of the vol across each time step is the same. This means that the time steps will be of unequal length.

Exercise 8.5 Modify the tree so the implied stock price process makes the discounted price a martingale. Compare convergence for calls, puts and forwards.

Exercise 8.6 Implement an American knock-in option pricer on a tree. (Use an additional auxiliary variable to indicate whether or not the option has knocked-in, and compute the value at each node in both cases.)

9

Solvers, templates and implied volatilities

9.1 The problem

Whatever model one is using to compute vanilla options' prices, it is traditional to quote prices in terms of the Black–Scholes implied volatility. The implied volatility is by definition the number to plug into the Black–Scholes formula to get the price desired. Thus we have the problem that we must solve the problem of finding the value σ such that

$$\text{BS}(S, K, r, d, T, \sigma) = \text{quoted price.}$$

In other words, we must invert the map

$$\sigma \mapsto \text{BS}(S, K, r, d, T, \sigma)$$

with the other parameters fixed.

The Black–Scholes formula is sufficiently complicated that there is no analytic inverse and this inversion must be carried out numerically. There are many algorithms for implementing such inversions; we shall study two of the simplest: bisection and Newton–Raphson. Our objective, as usual, is to illustrate the programming techniques for defining the interfaces in a reusable fashion rather than to implement the most efficient algorithms available. Indeed, we hope that the reader will combine the techniques here with algorithms found elsewhere, in for example [26], to produce robust and efficient reusable code.

Before proceeding to the coding and design issues, we recall the details of the aforementioned algorithms. Given a function, f, of one variable we wish to solve the equation

$$f(x) = y. \tag{9.1}$$

In the above, f is the Black–Scholes formula, x is volatility and y is the price. If

141

the function f is continuous, and for some a and b we have

$$f(a) < y, \tag{9.2}$$

$$f(b) > y, \tag{9.3}$$

then there must exist some c in the interval (a, b) such that $f(c) = x$. Bisection is one technique to find c. The idea is straightforward: we simply take the midpoint, m, of the interval, then one of three things must occur:

- $f(m) = y$ and we are done;
- $f(m) < y$ in which case there must be a solution in (m, b);
- $f(m) > y$ in which case there must be a solution in (a, m).

Thus by taking the midpoint, we either find the solution, or halve the size of the interval in which the solution exists. Repeating we must narrow in on the solution. In practice, we would terminate when we achieve

$$|f(m) - y| < \epsilon, \tag{9.4}$$

for some pre-decided tolerance, ϵ.

Bisection is robust but is not particularly fast. When we have a well-behaved function with an analytic derivative then Newton–Raphson can be much faster. The idea of Newton–Raphson is that we pretend the function is linear and look for the solution where the linear function predicts it to be. Thus we take a starting point, x_0, and approximate f by

$$g_0(x) = f(x_0) + (x - x_0)f'(x_0). \tag{9.5}$$

We have that $g_0(x)$ is equal to zero if and only if

$$x = \frac{y - f(x_0)}{f'(x_0)} + x_0. \tag{9.6}$$

We therefore take this value as our new guess x_1. We now repeat until we find that $f(x_n)$ is within ϵ of y.

Newton–Raphson is much faster than bisection provided we have an easily evaluated derivative. This is certainly the case for the Black–Scholes function. Indeed, for a call option the vega is easier to compute than the price is. However, as it involves passing two functions rather than one to a solver routine, it requires more sophisticated programming techniques to implement re-usably.

9.2 Function objects

We want to implement the bisection algorithm in a re-usable way; this means that we will need a way to pass the function f into the bisection routine. Since f may well be defined, as it is in our case, in terms of the value of a more complicated function with many parameters fixed, we will also need somehow to pass in the values of those auxiliary parameters.

There are, in fact, many different ways to tackle this problem. One method we have already studied is the engine template. With this approach, we define a base class for which the main method is to carry out the bisection. The main method calls a pure virtual method to get the value of $f(x)$. For any specific problem, we then define an inherited class which implements f appropriately. Whilst this method can work effectively, there are a couple of disadvantages. The first is that the function call is virtual which can lead to efficiency problems. There are two causes: the first is that to call a virtual function, the processor has to look up a virtual function table each time the function is called, and then jump to a location specified by the table. Clearly, it would be slightly faster not to have to look up the table. A more subtle and serious speed issue is that it is not possible to `inline` virtual functions. If the function is known beforehand, the compiler can inline it and eliminate mechanics of the function call altogether. In addition, the compiler may be able to make additional optimizations as it sees all the code together at once. Whilst these speed issues are not particularly important whilst designing a solver, they are more critical when writing a class for other numerical routines such as numeric integration where often a large of calls are made to a function which is fast to evaluate. We therefore wish to develop a pattern which can be used in those contexts too.

The second disadvantage of inheriting from a solver base class is that it inhibits other inheritance. If we wish to inherit the class defining our function from some other class, we cannot inherit from the solver class as well without using multiple inheritance. Of course, one could use multiple inheritance but it tends to be tricky to get it to work in a bug-free fashion and I therefore tend to avoid it.

Having decided that we want to be able to input a function to our routine without using virtual functions, what other options do we have? One solution would be to use a function pointer but this would buy us little (if anything) over virtual functions. Another approach is templatization. The crucial point is that with templatization the type of the function being used in the optimization is decided at compile time rather than at runtime. This means that the compiler can carry out optimizations and inlining that depend on the type of the function since that information is now available to it.

The approach we adopt for specifying the function we wish to optimize uses the *function object*. We first encountered function objects in Section 2.1 when defining pay-offs. Recall that a function object is by definition an object for which `operator()` is defined. So if we have an object f of a class T for which

```
const operator()( double x) const
```

has been defined it is then legitimate to write f(y) for a double y, and this is equivalent to

f.operator()(y).

Thus our object f can be used with function-like syntax. However, as f is an object it can contain extra information. Thus if we want to solve for the implied volatility, the function object will take the volatility as an argument, but will also have, as extra parameters already stored, the values of r, d, T, S and K.

We thus obtain the class defined in BSCallClass.h:

Listing 9.1

```
//
//  BSCallClass.h
//

#ifndef BS_CALL_CLASS_H
#define BS_CALL_CLASS_H
class BSCall
{

public:

    BSCall(double r_, double d_,
           double T, double Spot_,
           double Strike_);

    double operator()(double Vol) const;


private:

    double r;
    double d;
    double T;
    double Spot;
    double Strike;

};
#endif
```

The source file is simple:

Listing 9.2

```
//
//   BSCallClass.cpp
//

#include <BSCallClass.h>
#include <BlackScholesFormulas.h>

BSCall::BSCall(double r_, double d_,
               double T_, double Spot_,
               double Strike_)
               :
               r(r_),d(d_),
               T(T_),Spot(Spot_),
               Strike(Strike_)
{}

double BSCall::operator()(double Vol) const
{
    return BlackScholesCall(Spot,Strike,r,d,Vol,T);
}
```

The constructor simply initializes the class data members, which are the parameters needed to price a call option under Black–Scholes except the volatility. The operator() takes in the volatility and then invokes the Black–Scholes formula.

This is the simplest possible implementation of the class. If we were truly worried about efficiency considerations, we could code the formula directly and precompute as much of it as possible in the constructor. We could have for example a class data member, Moneyness, set to the log of Spot divided by Strike, and then we would not have to compute it every time.

9.3 Bisecting with a template

In the previous section, we showed how we could define a class for which the syntax f(x) makes sense when f was an object of the class, and x was a double. We still need to get the object f into our solver routine, however. We do so via templatization. The basic idea of templatization is that you can write code that works for many classes simultaneously provided they are required to have certain operations defined with the same syntax. In this case, our requirement is that the class should have

```
double operator()( double ) const
```

defined, and thus that the syntax f(y) is well-defined for class objects as we discussed above.

We present the Bisection function in Bisection.h:

Listing 9.3 (Bisection.h)

```
template<class T>
double Bisection(double Target,
                 double Low,
                 double High,
                 double Tolerance,
                 T TheFunction)
{
    double x=0.5*(Low+High);
    double y=TheFunction(x);

    do
    {
        if (y < Target)
            Low = x;

        if (y > Target)
            High = x;

        x = 0.5*(Low+High);

        y = TheFunction(x);
    }
    while
        ( (fabs(y-Target) > Tolerance) );

    return x;
}
```

We only present a header file, since for template code we cannot precompile in a source file – we do not know the type of the object T. The function is quite simple. We specify that it is templatized via the template<class T> at the top. If we invoke the function with the template argument BSCall via Bisection<BSCall> then every T will be converted into a BSCall before the function is compiled. Once we have fixed the type of the template argument, there is really very little to the function. We take the midpoint of the interval evaluate the function there, and switch to the left or right side of the interval by redefining Low and

High until the value at the midpoint is close enough to the target, and we then return.

Note that we have defined the type of the function object passed in as T The-Function: we could equally well have put const T& TheFunction. The syntax we have adopted involves copying the function object, and is therefore arguably less good than the alternative. The reason I have done it this way is to highlight the fact that the standard template library always uses the former syntax. A consequence of this is that one needs to be careful when using function objects with the STL not to define function objects which are expensive to copy (or, even worse, impossible to copy.)

We now give a simple example of an implied volatility function:

Listing 9.4 (SolveMain1.cpp)

```
/*
Needs
    BlackScholesFormulas.cpp
    BSCallClass.cpp
    Normals.cpp
*/
#include <Bisection.h>
#include <cmath>
#include <iostream>
#include <BSCallClass.h>
#include <BlackScholesFormulas.h>

using namespace std;

int main()
{
    double Expiry;
    double Strike;
    double Spot;
    double r;
    double d;
    double Price;

    cout << "\nEnter expiry\n";
    cin >> Expiry;

    cout << "\nStrike\n";
    cin >> Strike;
```

```
    cout << "\nEnter spot\n";
    cin >> Spot;

    cout << "\nEnter price\n";
    cin >> Price;

    cout << "\nr\n";
    cin >> r;

    cout << "\nd\n";
    cin >> d;

    double low,high;

    cout << "\nlower guess\n";
    cin >> low;

    cout << "\nhigh guess\n";
    cin >> high;

    double tolerance;

    cout << "\nTolerance\n";
    cin >> tolerance;

    BSCall theCall(r,d,Expiry,Spot,Strike);

    double vol =
        Bisection(Price,low,high,tolerance,theCall);
    double PriceTwo =
        BlackScholesCall(Spot,Strike,r,d,vol,Expiry);

    cout << "\n vol " << vol << " pricetwo "
                        << PriceTwo << "\n";

    double tmp;
    cin >> tmp;

    return 0;
}
```

As usual, we input all the necessary parameters. We then put them together to create a BSCall object. We then call the Bisection function to find the volatility. Note that we have not put Bisection<BSCall>. The compiler deduces the type of the template argument from the fact that our final argument in the function call is theCall. We only have to specify the template argument when the compiler does not have a sufficient amount of other information to deduce it.

Our function finishes by getting the price via the Black–Scholes functions for the implied volatility that was found. If everything is working correctly then this will give the original price inputted.

9.4 Newton–Raphson and function template arguments

We now want to adapt the pattern we have presented to work for Newton–Raphson as well as for bisection. The fundamental difference from a design viewpoint is that Newton–Raphson involves two functions, the value and the derivative, whereas bisection involves just one. One solution would be simply to pass in two function objects: one for the value and another for the derivative. This is unappealing, however, in that we would then need to initialize a set of parameters for each object and we would have to be careful to make sure they are the same. More fundamentally, the value and the derivative are really two aspects of the same thing rather than two separate functions and so having two objects does not express well our conceptual model of them.

A second solution is to assume a name for the derivative function. After all, that is essentially what we did for the value function; it was a special name with special syntax but ultimately it was just assuming a name. Thus we could assume that the class had a method

```
double Derivative(double ) const
```

defined and at appropriate points in our function we would then put

```
TheFunction.Derivative(x).
```

This would certainly work. However, it is a little ugly and if our class already had a derivative defined under a different name, it would be annoying.

Fortunately, there is a way of specifying which class member function to call at compile time using templatization. The key to this is a pointer to a member function. A pointer to a member function is similar in syntax and idea to a function pointer, but it is restricted to methods of a single class. The difference in syntax is that the class name with a :: must be attached to the * when it is declared. Thus to declare a function pointer called Derivative which must point to a method of the class T, we have

```
double (T::*Derivative)(double) const
```

The function `Derivative` is a `const` member function which takes in a `double` as argument and outputs a `double` as return value. If we have an object of class T called `TheObject` and `y` is a `double`, then the function pointed to can be invoked by

```
TheObject.*Derivative(y)
```

Whilst the syntax is a little cumbersome, the key is to realise that it is the same for ordinary function pointers except that `T::` must be added to the `*` for declarations and `TheObject.` must be added for invocations.

We can now use a function pointer to specify both the derivative and the value. As we would like to avoid the time spent on evaluating the pointers, we can make them template parameters rather than arguments to our function. This means that the compiler can treat them just like any other function call, and as their types are decided at compile time they can be `inlined`.

Our Newton–Raphson routine is therefore as follows:

Listing 9.5 (`NewtonRaphson.h`)

```
template<class T, double (T::*Value)(double) const,
                  double (T::*Derivative)(double) const >
                  double NewtonRaphson(double Target,
                                       double Start,
                                       double Tolerance,
                                       const T& TheObject)
{
    double y = (TheObject.*Value)(Start);

    double x=Start;

    while ( fabs(y - Target) > Tolerance )
    {
        double d = (TheObject.*Derivative)(x);

        x+= (Target-y)/d;

        y = (TheObject.*Value)(x);
    }
    return x;
}
```

We have three template parameters: the class, the pointer to the value function for that class, and the pointer to the derivative function for that class. The routine is short and simple, now that we have the right structure. As usual, we keep repeating until close enough to the root. We have not included the checks required to ensure that the loop does not repeat endlessly if the sequence fails to converge to a root, but such additions are easily put in.

9.5 Using Newton–Raphson to do implied volatilities

Now that we have developed a Newton–Raphson routine, we want to use it to compute implied volatilities. Our class will therefore have to support pricing as a function of volatility and the vega as a function of volatility. As before, the other parameters will be class data members which are not inputted in the constructor rather than via these methods. We present a suitable class in BSCallTwo.h and BSCallTwo.cpp.

Listing 9.6 (BSCallTwo.h)

```
#ifndef BS_CALL_TWO_H
#define BS_CALL_TWO_H

class BSCallTwo
{

public:
    BSCallTwo(double r_, double d_,
                    double T, double Spot_,
                    double Strike_);

    double Price(double Vol) const;
    double Vega(double Vol) const;

private:
    double r;
    double d;
    double T;
    double Spot;
    double Strike;
};
#endif
```

The methods just call the relevant functions. As before, we could optimize by pre-computing as much as possible, and inlining the methods.

Listing 9.7 (BSCallTwo.cpp)

```cpp
#include <BSCallTwo.h>
#include <BlackScholesFormulas.h>


BSCallTwo::BSCallTwo(double r_, double d_,
                     double T_, double Spot_,
                     double Strike_)
                     :
                     r(r_),d(d_),
                     T(T_),Spot(Spot_),
                     Strike(Strike_)
{}


double BSCallTwo::Price(double Vol) const
{
    return BlackScholesCall(Spot,Strike,r,d,Vol,T);
}


double BSCallTwo::Vega(double Vol) const
{
    return BlackScholesCallVega(Spot,Strike,r,d,Vol,T);
}
```

We present an example of using the combination of NewtonRaphson and BSCallTwo in SolveMain2.cpp.

Listing 9.8 (SolveMain2.cpp)

```cpp
/*
Needs
    BlackScholesFormulas.cpp
    BSCallTwo.cpp
    Normals.cpp
*/
#include <NewtonRaphson.h>
#include <cmath>
#include <iostream>
#include <BSCallTwo.h>
#include <BlackScholesFormulas.h>
```

```
using namespace std;

int main()
{
    double Expiry;
    double Strike;
    double Spot;
    double r;
    double d;
    double Price;

    cout << "\nEnter expiry\n";
    cin >> Expiry;

    cout << "\nStrike\n";
    cin >> Strike;

    cout << "\nEnter spot\n";
    cin >> Spot;

    cout << "\nEnter price\n";
    cin >> Price;

    cout << "\nr\n";
    cin >> r;

    cout << "\nd\n";
    cin >> d;

    double start;

    cout << "\nstart guess\n";
    cin >> start;

    double tolerance;

    cout << "\nTolerance\n";
    cin >> tolerance;

    BSCallTwo theCall(r,d,Expiry,Spot,Strike);
```

```
double vol=NewtonRaphson<BSCallTwo, &BSCallTwo::Price,
                   &BSCallTwo::Vega>(Price, start,
                                          tolerance, theCall);

double PriceTwo =
    BlackScholesCall(Spot,Strike,r,d,vol,Expiry);

cout << "\n vol " << vol << " \nprice two:" << PriceTwo
                                    << "\n";

double tmp;
cin >> tmp;

return 0;
}
```

Our new main program is very similar to the one we had before. The main change is that this time we specify the template parameters for our NewtonRaphson function, whereas for the Bisection function we did not bother. The reason for the change is that there is not sufficient information for the compiler to deduce the types of the parameters. There is nothing to indicate which member functions of the class are to be used. Even for our class which only has two member functions, these two functions could equally well be the other way round as far the compiler knows.

9.6 The pros and cons of templatization

In this chapter, we have used template arguments to achieve re-usability whereas in other chapters, we have used virtual functions and polymorphism. There are advantages and disadvantages to each approach. The principal difference is that for templates argument types are decided at the time of compilation, whereas for virtual functions the type is not determined until runtime.

What consequences does this difference have? The first is speed. No time is spent on deciding which code to run when the code is actually running. In addition, the fact that the compiler knows which code will be run allows it to make extra optimizations which would be hard if not impossible when the decision is made at run time.

A second consequence is size. As the code is compiled for each template argument used separately, we have multiple copies of very similar code. For a simple routine such as a solver this is not really an issue but for a complicated routine, this could result in a very large executable. Another aspect of this is slower compiler times since much more code would have to be compiled. If we had several template parameters the size could multiply out of control. For example, suppose when designing our Monte Carlo path-dependent exotic pricer, we had templatized both the random number generator and the product. If we had six random number generators and ten products, and we wished to allow any combination then we would have sixty times as much code.

A third consequence is that it becomes harder for the user of the code to make choices. In the example of the exotics pricer, if the user was allowed to choose the number generator and the product via outside input, we would have to write code that branched into each of the sixty cases and within each branch called the engine and gathered results.

As well as the run time versus compile time decision, there are other issues with using templatized code. A simple one is that it is harder to debug. Some debuggers get confused by template code and, for example, refuse to set breakpoints within templates (or else set and ignore them.) Related to this is the fact that compilers will often not actually compile lines of template code that are not used. Thus if a templatized class has a method that is not called anywhere, the code will compile even if it has syntax errors. Only when a line is added that calls the particular method will the compiler errors appear. This can be infuriating as the error may show up a long time afterwards.

One way of avoiding these problems is first to write non-template code for a particular choice of the template parameter. This code can be thoroughly tested and debugged, and then afterwards the code can be rewritten by changing the particular parameter into a template parameter.

So when should one use templates and when use virtual functions? My preference is not to use templates unless certain conditions are met. These are that the routine should be short, and potentially re-usable in totally unrelated contexts. So for example, I would use templates for a numerical integration routine and a solver. I would also use templates for a container class; in fact I would generally use the templates given in the standard template library. I would not, however, use templates for an option pricing engine since the code would probably be long and is only relevant to a quite specific context.

The general trend in C++ is towards templates. The principal reason is that they are the way to achieve the same speed as lower-level languages. In general, languages exhibit a trade-off between abstraction and efficiency; C++ has always striven to achieve both. Templates are ultimately a way of achieving abstraction without sacrificing efficiency.

9.7 Key points

In this chapter we have looked at how to implement solvers using template code.

- Templates are an alternative to inheritance for coding without knowing an object's precise type.
- Template code can be faster as function calls are determined at compile time.
- Extensive use of template code can lead to very large executables.
- Pointers to member functions can be a useful way of obtaining generic behaviour.
- Implied volatility can only be computed numerically.

9.8 Exercises

Exercise 9.1 Modify the Newton–Raphson routine so that it does not endlessly loop if a root is not found.

Exercise 9.2 Take your favourite numerical integration routine, e.g. the trapezium rule, and write a template routine to carry it out.

Exercise 9.3 Write a routine to price a vanilla option by Monte Carlo or trees where the pay-off is passed in as a template parameter expressed via a function object.

10

The factory

10.1 The problem

Suppose we wish to design an interface which is a little more sophisticated than those we have used so far. The user will input the name of a pay-off and a strike, and the program will then price a vanilla option with that pay-off. We therefore need a conversion routine from strings and strikes to pay-offs. How might we write this?

One simple solution is to write a function that takes in the string and the strike, checks against all known types of pay-offs and when it comes across the right one, creates a pay-off of the right type. We would probably implement this via a switch statement. Our conversion routine would then have to include the header files for all possible forms of pay-off, and every time we added a new pay-off we would have to modify the switch statement. Clearly, this solution violates the open-closed principle as any addition involves modification.

In this chapter, we present a solution that allows us to add new pay-offs without changing any of the existing files. We simply add new files to the project. Our solution is a design pattern known as *the factory pattern*. It is so called because it can be used to manufacture objects. Whilst we restrict attention to a simple factory which manufactures pay-offs, the basic pattern can be used in much wider contexts.

10.2 The basic idea

Our solution requires each type of pay-off to tell the factory that it exists, and to give the factory a blueprint for its manufacture. In this context a blueprint means an identity string to distinguish that class and a pointer to a function that will create objects of that class.

How can we get the class to communicate with the factory, without explicitly calling anything from the main routine? The key lies in global variables. Every global variable is initialized when the program commences before anything else

157

happens. If we define a class in such a way that initializing a global variable of that class registers a pay-off class with the factory, then we have achieved what we wanted. This is possible because the initialization involves a call to a constructor, and we can make the constructor do whatever we want.

So for each pay-off class, we write an auxiliary class whose constructor registers the pay-off class with our factory, and we declare a global variable of the auxiliary class. In fact, as these auxiliary classes will all be very similar to each other, we adopt a template solution for defining these classes.

We also need a factory object for these auxiliary classes to talk to. We cannot make this factory object a global variable, as we have no control over the order in which the global variables are initialized. We need it to exist before the other globals are defined as they refer to it. Fortunately, there is a type of variable guaranteed to come into existence at the moment it is first referred to: the `static` variable.

Thus if the registration function contains a `static` variable which is the factory, then on the first call to the registration function the factory comes into existence. Recall that a `static` variable defined in a function persists from one call to the next, and only disappears when the program exits. So all the registration function calls will register the pay-off blueprints with the same factory.

However, we are not yet done because the creator function will need to have access to the same factory object as the registration function, and if the factory is hidden inside the registration function this will not be possible. The solution to this problem is known as the *singleton* pattern.

10.3 The singleton pattern

We saw in the last section that we need to define a factory via a `static` variable since it must come into existence as soon as it is referred to when registering the blueprints. We also saw that the same factory must be referred to by every registration, and that the same factory will be needed when creating the pay-offs from strings.

So what we need is a factory class for which an object exists as soon as it is required, and for this object to exist until the end of the program. We also do not want any other factory objects to exist as they will just confuse matters; everything must be registered with and built by the same factory.

The singleton pattern gives a way of creating a class with these properties. The first thing is that all constructors and assignment operators are made private. This means that factories can only be created from inside methods of the class this gives us firm control over the existence of factory objects. In order to get the one class object that we need, we define a very simple method that defines a class object as a static variable. Thus if our class is called `PayOffFactory`, we define a class method `Instance` as follows:

```
PayOffFactory& PayOffFactory::Instance()
{
    static PayOffFactory theFactory;
    return theFactory;
}
```

The first time that `Instance` is called, it creates the `static` data member `the-Factory`. As it is a member function, it can do this by using the private default constructor. Every subsequent time the `Instance` is called, the address of the already-existing `static` variable `theFactory` is returned. Thus `Instance` creates precisely one `PayOffFactory` object which can be accessed from anywhere by calling `PayOffFactory::Instance()`.

Note that `Instance` will have to be a `static` method of `PayOffFactory`, as the whole point is that it provides you with a `PayOffFactory` object, and it would be useless if you had to access it from an existing object. Note also that the meaning of `static` here for a *function* is quite different from the meaning above for a *variable*; for a function it means that the function can be called directly without any attachment to an object. One still has to prefix with the name of class, however.

We have now achieved what we needed: we have a way of creating a single factory which can be referenced from anywhere at any time in the program. Note that the name singleton pattern was chosen because precisely one object from the class exists.

10.4 Coding the factory

In the last section, we saw how the singleton pattern could be used to create a single factory accessible in any place at any time. We now use this pattern to implement the factory. As well as the instance method discussed above, we will need a method for registering pay-off classes and a method for creating then.

How will registration work? Upon registration, we need to know the string identifier for the specific pay-off class and the pointer to the function which actually creates the object in question. These will therefore be the arguments for the registration method. The factory will need to store this information for when the create pay-off method is called. This will require a container class.

Fortunately, there is a container in the standard template library which is designed for associating identifiers to objects. This container is called the `map` class. We therefore need a data member which is a map with template arguments `std::string` and pointers to create functions.

Finally, we need a method which turns a string plus a strike into a `PayOff` object. Our header file is therefore as follows:

Listing 10.1 (PayOffFactory.h)

```
#ifndef PAYOFF_FACTORY_H
#define PAYOFF_FACTORY_H
#include <PayOff3.h>

#if defined(_MSC_VER)
#pragma warning( disable : 4786)
#endif

#include <map>
#include <string>

class PayOffFactory
{
public:
    typedef PayOff* (*CreatePayOffFunction)(double );

    static PayOffFactory& Instance();
    void RegisterPayOff(std::string, CreatePayOffFunction);
    PayOff* CreatePayOff(std::string PayOffId,double Strike);
    ~PayOffFactory(){};

private:
    std::map<std::string, CreatePayOffFunction>
                TheCreatorFunctions;
    PayOffFactory(){}
    PayOffFactory(const PayOffFactory&){}
    PayOffFactory& operator=
                (const PayOffFactory&){ return *this;}
};
#endif
```

Note the typedef: this allows us to refer to pointers to functions which take in a double and spit out a PayOff* as CreatePayOffFunction. Without this typedef, the syntax would quickly become unmanageable. Note also that we make the CreatePayOff method return a pointer to a PayOff. The reason for this is that it allows the possibility of returning a null pointer if the identity string was not found; otherwise we would have to throw an error or return a default sort of pay-off.

   We present the source code in PayOffFactory.cpp:

Listing 10.2 (`PayOffFactory.cpp`)

```cpp
#if defined(_MSC_VER)
#pragma warning( disable : 4786)
#endif

#include <PayOffFactory.h>
#include <iostream>
using namespace std;

void PayOffFactory::RegisterPayOff(string PayOffId,
                        CreatePayOffFunction CreatorFunction)
{
 TheCreatorFunctions.insert(pair<string,CreatePayOffFunction>
                            (PayOffId,CreatorFunction));
}


PayOff* PayOffFactory::CreatePayOff(string PayOffId,
                                    double Strike)
{
 map<string, CreatePayOffFunction>::const_iterator
                    i = TheCreatorFunctions.find(PayOffId);

 if  (i == TheCreatorFunctions.end())
 {
     std::cout << PayOffId
     << " is an unknown payoff" << std::endl;
     return NULL;
 }

 return (i->second)(Strike);
}

PayOffFactory& PayOffFactory::Instance()
{
  static PayOffFactory theFactory;
  return theFactory;
}
```

Other than for the `Instance()` method, which we have already discussed, the methods are really just wrappers for the inner map object.

In case the reader is not familiar with the STL map container, we discuss a little how it works. A map is a collection of pairs. We used the pair class to store the nodes of a tree in Chapter 8. Recall that a `pair` is a simple class consisting of two public data members known as `first` and `second`. The types of these data members are template parameters. When working with a map, `first` is the key or identifier used to look up the object we wish to find which is stored in `second`.

For us this means that the type of the map is

```
map<std::string, CreatePayOffFunction>
```

and every `pair` that we use will be of the same type. The `insert` method is used to place `pairs` of `strings` and `CreatePayOffFunctions` into the `map`. A map has the property that each key is unique so if you insert two `pairs` with the same key then the second one is ignored. For us, this means that if we give two `PayOff` classes the same string identifier only one will be registered. It is possible to examine the return type of the `insert` to determine whether the insertion was successful. The method `RegisterPayOff` carries out this insertion for our factory.

The retrieval is carried out in `CreatePayOff`: a `string` is passed in and the `find` method of `map` is used. This method returns a `const_iterator` pointing to the `pair` which has the correct key (i.e. `first` element) if such a `pair` exists, and otherwise the iterator points to the `end` of the `map`. For the reader who is not familiar with iterators, an iterator is an abstraction of a pointer and works in similar fashion. Just like pointers, they can be dereferenced via `*` or `->`. A `const_iterator` is similar to a non-`const` pointer to `const` objects. That is, the iterator's value can be changed, but the value of the thing it points to cannot.

Our method therefore uses `find` to get an iterator. This iterator is then checked to see if the look-up succeeded. If it failed we print an error message and return a null pointer. If is succeeded, we take the `second` element of the `pair` pointed to, which is a function pointer, dereference it and call it with argument `Strike`. Since this function's job is to create a `PayOff` object of the relevant type and return a pointer to it, we have achieved our objective; objects of any class previously registered can be created by entering the appropriate string and strike.

Whilst we certainly could have programmed our factory without using the STL `map` class, its existence certainly made the task much easier. We refer the reader to Stroustrup, [29], Section 17.4, Josuttis, [11], and Meyers, [18], for further information on the `map` class.

10.5 Automatic registration

We discussed above how we could manage registration of `PayOff` classes by using global variables. Here we look at how to carry out this out. As we mentioned above, the code is the same except for class names for each registration so it makes sense to use a template class. We present this template code in `PayOffConstructible.h`:

Listing 10.3 (`PayOffConstructible.h`)

```
#ifndef PAYOFF_CONSTRUCTIBLE_H
#define PAYOFF_CONSTRUCTIBLE_H

#if defined(_MSC_VER)
#pragma warning( disable : 4786)
#endif

#include <iostream>
#include <PayOff3.h>
#include <PayOffFactory.h>
#include <string>

template <class T>
class PayOffHelper
{
public:
    PayOffHelper(std::string);
    static PayOff* Create(double);
};

template <class T>
PayOff* PayOffHelper<T>::Create(double Strike)
{
    return new T(Strike);
}

template <class T>
PayOffHelper<T>::PayOffHelper(std::string id)
{
    PayOffFactory& thePayOffFactory = PayOffFactory::Instance();
    thePayOffFactory.RegisterPayOff(id,PayOffHelper<T>::Create);
}
#endif
```

The helper class we define here has to do two things. It must define a constructor that carries out the registration of the class defined by the template parameters, and it must define a function which will carry out the creation so we have something to use in the registration process!

The constructor takes in a `string` as an argument; this string will be needed to identify the class being registered. The constructor simply first calls `Instance` to get the address of the factory object, and then calls the `RegisterPayOff` method of the factory to carry out the registration. Note that the constructor does not actually do anything as regards the actual object being created! In fact, the class has no data members so it would not be possible to do anything.

The method `Create` defines the function used to create the pay-off object on demand. Note that it is `static` as it should not be associated to any particular class object. The function simply calls the constructor for objects of type `T` with argument `Strike`. Of course, there is something slightly subtle here in that the specification of the template parameter, `T`, is making the choice of which object to construct. Note that we use `new` as we want the created object to persist after the function is finished. One consequence of this is that the object will have to be properly `deleted` at some point.

`PayOffRegistration.cpp` includes an example of using the `PayOff Helper` class.

Listing 10.4 (`PayOffRegistration.cpp`)

```
#include <PayOffConstructible.h>

namespace
{
  PayOffHelper<PayOffCall> RegisterCall("call");

  PayOffHelper<PayOffPut> RegisterPut("put");
}
```

Note that if we were defining a new class, we would probably put this registration in the source file for the class but as we have already defined the call and put classes, we do not do so here. The registration file is quite short. We define two global variables, `RegisterCall` and `RegisterPut`. These are of type `PayOffHelper<Call>` and `PayOffHelper<Put>`. As global variables, they are initialized at the start of the program. This initialization carries out the registration as required. Note that we have put a `namespace` command around the declaration. This means that the variables are in an unnamed namespace and as such are invisible to the rest of the program. So the variables are both global and invisible. Why do we want them invisible? Their purpose is purely to perform the registration, and once that has been done we have no further use for them so it is best to put them out of sight and out of temptation's reach.

10.6 Using the factory

Now we have done all the set-up work, how do we use the factory? We give a very simple example in `PayFactoryMain.cpp`.

Listing 10.5 (`PayFactoryMain.cpp`)

```
/*
Uses
    PayOff3.cpp
    PayOffBridge.cpp
    PayOffFactory.cpp
    PayOffRegistration.cpp
*/


#include <PayOff3.h>
#include <PayOffConstructible.h>
#include <PayOffBridge.h>
#include <PayOffFactory.h>
#include <string>
#include <iostream>
using namespace std;

int main()
{
    double Strike;
    std::string name;

    cout << "Enter strike\n";
    cin >> Strike;

    cout << "\npay-off name\n";
    cin >> name;

    PayOff* PayOffPtr =
        PayOffFactory::Instance().CreatePayOff(name,Strike);

    if (PayOffPtr != NULL)
    {
        double Spot;

        cout << "\nspot\n";
        cin >> Spot;
```

```
        cout << "\n" << PayOffPtr->operator ()(Spot) << "\n";
        delete PayOffPtr;
    }

    double tmp;
    cin >> tmp;
    return 0;
}
```

This routine is very simple but illustrates the important points. The user inputs spot, strike and the name of the option. If an option with that name has been registered then the pay-off is computed, and the object is then deleted.

  The important point here is that the name definitions are carried out in the file `PayOffRegistration.cpp`, and this file is not seen directly by any of the other files including the `main` routine. If we wanted to add another `PayOff`, say the forward, we could so without modifying any of the existing files. In fact, all we would have to do is add the header and source file for the forward, and in a new file `PayOffForwardRegistration.cpp` add the declaration

`PayOffHelper<PayOffForward> RegisterForward("forward");`

As we originally required, this would not require recompilation of any of the original files. We have therefore achieved our original objective of an open-closed pattern.

10.7 Key points

In this chapter we have developed the factory pattern and the singleton pattern in order to give a method of adding new pay-off classes to an interface without modifying existing files.

- The singleton pattern allows us to create a unique global object from a class and provide a way of accessing it.
- The factory pattern allows us to add extra inherited classes to be accessed from an interface without changing any existing files.
- The factory pattern can be implemented using the singleton pattern.
- The standard template library `map` class is a convenient way to associate objects with string identifiers.
- Placing objects in an unnamed namespace is a way of ensuring that they are not accessed elsewhere.
- We can achieve automatic registration of classes by making their registration a side-effect of the creation of global variables from a helper class.

10.8 Exercises

Exercise 10.1 Write a straddle class and register it with the factory.

Exercise 10.2 Our class cannot handle a double digital as it needs two strikes. Work out a solution that will handle options with multiple parameters.

Exercise 10.3 Integrate the factory with a Monte Carlo routine.

11

Design patterns revisited

11.1 Introduction

In this final chapter, we revisit and catalogue the design patterns from earlier chapters. We also mention a few other patterns we have not studied which the reader may find helpful. Finally, we discuss further reading on the topic of design patterns.

The design patterns we have studied are a small subset of those in the classic book on the topic: *Design Patterns*, [6], which is often referred to as the 'Gang of Four' book. As well as listing the patterns, the authors attempt to classify the patterns according to the contexts in which they are used. In this final chapter, we revisit the patterns the we have studied in the context of that classification. We also mention some patterns discussed there which we have not examined here.

11.2 Creational patterns

A creational pattern is a pattern that deals primarily with the creation of new objects. Their purpose is to abstract the creation process which helps the system to be developed independently of the types of individual objects. In fact, sometimes all we can be sure of about these objects, is what class they are inherited from, or in other terms what interface they implement.

11.2.1 Virtual copy constructor

We have extensively used the concept of cloning. We need a copy of an object, we do not know its type so we cannot use the copy constructor so we ask the object to provide a copy of itself. Note the general philosophy here is that the object knows more about itself than we do so we ask it to help us out. Note we could easily modify this pattern to ask the object to make a default object from its class, as once again it knows its class type and we do not. In [6] virtual constructors are known as the 'Factory Method'.

11.2.2 The factory

This is called the 'abstract factory' in [6]. The purpose of this pattern is to allow us to have an object that spits out objects as and when we need them. This means in particular that responsibility for creating objects of the relevant type lies with a single object. We thus gain greater control over their creation, which yields greater flexibility in changing the objects used. We principally used this pattern to give an easily extended interface. In particular, we saw that the pattern allows the addition of new classes to an interface without the rewriting of any code.

11.2.3 Singleton

We used the singleton pattern to implement our factory. The big advantage of the singleton pattern is that there is a single copy of the object which is accessible from everywhere, without introducing global variables and all the difficulties they imply. Note that if for some reason we wanted more than one copy of the object to exist then we could easily modify the pattern to given us a doubleton or a tripleton and so on. For example, we could define more than one method that performed similarly to our `Instance` method.

11.2.4 Monostate

We have not examined the monostate pattern nor is it covered in *Design Patterns*, however, it is a useful alternative to the singleton pattern. Rather than only allowing one object from the class to exist, we allow an unlimited number but make them all have the same member variables. Thus all the objects from the class act as one. The way we do this is by making all the data members `static`. This approach allows us to treat each object from the class like any other, although they are all really the same object. Our factory could easily have been implemented using this pattern.

11.3 Structural patterns

A structural pattern is one that deals mainly with how classes are composed to define more intricate structures. They allow us to design code that has extra functionality without having to rewrite existing code.

11.3.1 Adapter

The adapter is a class that translates an interface into a form that other classes expect. It is most useful when we wish to fit code into a structure for which it was not originally designed, either because we have changed our way of doing things or because the code originates elsewhere. For example, if we download a library from the web, its interface is unlikely to conform to what we have been using. By adapting the interface, we can seamlessly integrate it into existing code. We gave an example of the adapter when implementing random number generators.

11.3.2 Bridge

The bridge is similar to the adapter in that it defines an interface, and acts as an intermediary between a client class and the classes implementing the interface. Thus the implementing class can easily be changed without the client class being aware of the change. The main difference between the bridge and the adapter is that the bridge is intended to define an intermediary interface from the start, whereas the adapter is introduced a later stage in order to solve incompatibilities. We used the bridge to create the `PayOff` and `Parameters` objects.

11.3.3 Decorator

The decorator patterns allows us to change the behaviour of a class at run-time without changing its interface. We add a wrapper class that processes incoming or outgoing messages and then passes them on. We saw that a decorator could be used to implement anti-thetic sampling when studying random number generation. We also saw that it could be used to create convergence tables of statistics when designing statistics gatherers. An attractive aspect of decoration is that we can decorate as many times as we like since the interface after decoration is the same as the interface before.

11.4 Behavioural patterns

Behavioural patterns are used for the implementation of algorithms. They allow us to vary aspects of algorithms interchangeably. They can also allow us to re-use algorithms in wildly unrelated contexts.

11.4.1 Strategy

In the strategy pattern, we defer an important part of algorithm to an inputted object. This allows us to easily change how this particular part of the algorithm behaves. We have used this pattern implicitly and explicitly all through the book. By making the `PayOff` or `VanillaOption` an input to our pricer, we are using this pattern. Less trivially, we made the random number generator an input to our Monte Carlo, and the generator is a key part of the algorithm.

11.4.2 Template

In the template pattern, rather than inputting an aspect of the algorithm, we defer part of the algorithm's implementation to an inherited class. The base class thus provides the structure of how the different parts of the algorithm fit together, but does not specify all the details of the implementation. We adopted this approach when designing our exotics Monte Carlo pricer; there we defined the process for the stock price evolution in an inherited class. This allowed the possibility of pricing exotics using a different model in the future.

11.4.3 Iterator

We have only briefly mentioned iterators; however, they are an important component of the standard template library. An iterator is essentially an abstraction of a pointer. As such it should be possible to dereference it, i.e. look at what it points to, increment and decrement it. The idea of iterators is that one can be defined for any sort of data structure, and so if an algorithm is defined in terms of iterators it can be applied to any sort of data structure. In the STL, algorithms take the type of the iterator as a template argument, which allows this generality to be implemented.

11.5 Why design patterns?

Why think in terms of design patterns? What has this classification bought us? The first simple thing is that it becomes much easier to explain our code to someone else; remember that re-use is ultimately defined socially not analytically. When describing our code to someone else, if we can describe a class by saying this is such and such standard pattern then they immediately have a mental model of how it works from their previous familiarity with that pattern.

A second advantage is that by having familarity with a collection of standard design patterns, we gain an immediate toolbox for solving any problem put in front of us. Thus when confronted with a programming problem, we can approach it by thinking "what design pattern is appropriate here?" rather than by attempting to solve it from scratch. Even if none of the known patterns are appropriate, examination of the problem through the lens of design patterns will help us to solve it.

In particular, the knowledge of why the patterns are inappropriate will aid us in developing a solution.

Of course, most programmers have patterns they implicitly use regularly. Indeed, many experienced programmers reading this book may feel they are only learning a formalization of what they did in any case. The advantage for them in using patterns is that it will help them to think more clearly about what they do and why.

11.6 Further reading

There are now many books on the topic of design patterns. We mention a few that the author has found useful. One good and straightforward book which was deliberately written as an easier companion to *Design Patterns* is *Design Patterns Explained* by Shalloway & Trott. The authors carefully go through many patterns explaining in simple language the concepts introduced in the original book.

C++ Programming: with Design Patterns Revealed by Muldner is another introductory book which is accessible, and takes the point of view that C++ should be learnt from the start in terms of design patterns.

Modern C++ Design by Alexandrescu is a more advanced book. It covers many more intricate ideas using templates than we have had the opportunity to cover here.

As well as books explicitly on design patterns, a C++ programmer needs many other standard texts. Some favourites of the author are

Effective C++, *More Effective C++* and *Effective STL* by Scott Meyers. These books are collections of programming gems by a C++ expert.

The C++ Programming Language by Bjarne Stroustrup. This is the ultimate reference book on C++ by the man who invented the language.

The C++ Standard Library by Nicolai Josuttis. This is a comprehensive description of the standard library that ships with any C++ compiler.

My favourite introductory book on object-oriented programming is

The Tao of Objects by Gary Entsminger. It's an easy read and concentrates on introducing the basic ideas of OO design.

11.7 Key points

- Design patterns can be classified into behavioural, structural and creational patterns.
- Behavioural patterns are used for the implementation of algorithms.

- Structural patterns deal with how classes are composed to create more intricate designs.
- A creational pattern deals primarily with the creation of new objects.

11.8 Exercise

Exercise 11.1 Implement the factory from Chapter 10 using the monostate pattern instead of the singleton pattern.

Appendix A

Black–Scholes formulas

In Chapter 9, we developed an implied volatility function; this function necessarily meant we needed formulas for the Black–Scholes price and vega of a call option. We present code for the pricing of calls, puts, digital calls and digital puts, and for the vega of a call option in BlackScholesFormulas.h and BlackScholes-Formulas.cpp. We do not present the formulas here as they can be found in just about any book on derivatives pricing, e.g. [12]. The code is a straightforward implementation of formulas. Note that we rely heavily on having an implementation of the cumulative normal function. We give such an implementation in Appendix B.

Listing A.1 (BlackScholesFormulas.h)

```
#ifndef BLACK_SCHOLES_FORMULAS_H
#define BLACK_SCHOLES_FORMULAS_H

double BlackScholesCall( double Spot,
                         double Strike,
                         double r,
                         double d,
                         double Vol,
                         double Expiry);

double BlackScholesPut( double Spot,
                        double Strike,
                        double r,
                        double d,
                        double Vol,
                        double Expiry);
```

```
double BlackScholesDigitalCall(double Spot,
                               double Strike,
                               double r,
                               double d,
                               double Vol,
                               double Expiry);


double BlackScholesCallVega( double Spot,
                             double Strike,
                             double r,
                             double d,
                             double Vol,
                             double Expiry);


#endif
```

Listing A.2 (BlackScholesFormulas.cpp)

```
#include <BlackScholesFormulas.h>
#include <Normals.h>
#include <cmath>

#if !defined(_MSC_VER)
using namespace std;
#endif

double BlackScholesCall( double Spot,
                         double Strike,
                         double r,
                         double d,
                         double Vol,
                         double Expiry)
{
  double standardDeviation = Vol*sqrt(Expiry);
  double moneyness = log(Spot/Strike);
  double d1 =( moneyness + (r-d)*Expiry +
  0.5* standardDeviation*standardDeviation)/standardDeviation;
  double d2 = d1 - standardDeviation;
  return Spot*exp(-d*Expiry) * CumulativeNormal(d1) -
   Strike*exp(-r*Expiry)*CumulativeNormal(d2);
}
```

```
double BlackScholesPut( double Spot,
                        double Strike,
                        double r,
                        double d,
                        double Vol,
                        double Expiry)
{
  double standardDeviation = Vol*sqrt(Expiry);
  double moneyness = log(Spot/Strike);
  double d1 =( moneyness + (r-d)*Expiry +
  0.5* standardDeviation*standardDeviation)/standardDeviation;
  double d2 = d1 - standardDeviation;
  return Strike*exp(-r*Expiry)*(1.0-CumulativeNormal(d2)) -
    Spot*exp(-d*Expiry) * (1-CumulativeNormal(d1));
}


double BlackScholesDigitalCall( double Spot,
                                double Strike,
                                double r,
                                double d,
                                double Vol,
                                double Expiry)
{
  double standardDeviation = Vol*sqrt(Expiry);
  double moneyness = log(Spot/Strike);
  double d2 =( moneyness + (r-d)*Expiry -
  0.5* standardDeviation*standardDeviation)/standardDeviation;
  return exp(-r*Expiry)*CumulativeNormal(d2);
}


double BlackScholesDigitalPut( double Spot,
                               double Strike,
                               double r,
                               double d,
                               double Vol,
                               double Expiry)
{
  double standardDeviation = Vol*sqrt(Expiry);
  double moneyness = log(Spot/Strike);
  double d2 =( moneyness + (r-d)*Expiry -
```

```
    0.5* standardDeviation*standardDeviation)/standardDeviation;
    return exp(-r*Expiry)*(1.0-CumulativeNormal(d2));
}


double BlackScholesCallVega( double Spot,
                             double Strike,
                             double r,
                             double d,
                             double Vol,
                             double Expiry)
{
    double standardDeviation = Vol*sqrt(Expiry);
    double moneyness = log(Spot/Strike);
    double d1 =( moneyness +  (r-d)*Expiry +
    0.5* standardDeviation*standardDeviation)/standardDeviation;
    return Spot*exp(-d*Expiry) * sqrt(Expiry)*NormalDensity(d1);
}
```

Appendix B

Distribution functions

We have repeatedly used the cumulative normal distribution function for a standard normal random variable and its inverse. We give an implementation of rational approximations for these functions in `Normals.h` and `Normals.cpp`. For further discussion of these approximations see [4], [19] and [9].

Listing B.1 (`Normals.h`)

```
#ifndef NORMALS_H
#define NORMALS_H


double NormalDensity(double x);


double CumulativeNormal(double x);


double InverseCumulativeNormal(double x);


#endif
```

Listing B.2 (`Normals.cpp`)

```
/*
code to implement the basic distribution functions necessary
inmathematical finance via rational approximations
*/


#include <cmath>
#include <Normals.h>


// the basic math functions should be in namespace std but
//aren't in VCPP6
```

```cpp
#if !defined(_MSC_VER)
using namespace std;
#endif


const double OneOverRootTwoPi = 0.398942280401433;


// probability density for a standard Gaussian distribution
double NormalDensity(double x)
{
    return OneOverRootTwoPi*exp(-x*x/2);
}


// the InverseCumulativeNormal function via the
// Beasley-Springer/Moro approximation
double InverseCumulativeNormal(double u)
{

    static double a[4]={  2.50662823884,
                        -18.61500062529,
                         41.39119773534,
                        -25.44106049637};

    static double b[4]={-8.47351093090,
                        23.08336743743,
                       -21.06224101826,
                         3.13082909833};

    static double c[9]={0.3374754822726147,
                        0.9761690190917186,
                        0.1607979714918209,
                        0.0276438810333863,
                        0.0038405729373609,
                        0.0003951896511919,
                        0.0000321767881768,
                        0.0000002888167364,
                        0.0000003960315187};

    double x=u-0.5;
    double r;
```

```
if (fabs(x)<0.42) // Beasley-Springer
{
    double y=x*x;

    r=x*(((a[3]*y+a[2])*y+a[1])*y+a[0])/
            (((((b[3]*y+b[2])*y+b[1])*y+b[0])*y+1.0));

}
else // Moro
{

    r=u;

    if (x>0.0)
        r=1.0-u;

    r=log(-log(r));

    r=c[0]+r*(c[1]+r*(c[2]+r*(c[3]+r*(c[4]+r*(c[5]+
            r*(c[6]+r*(c[7]+r*c[8]))))))));

    if (x<0.0)
        r=-r;

}

    return r;
}


// standard normal cumulative distribution function
double CumulativeNormal(double x)
{
    static double a[5] = { 0.319381530,
                          -0.356563782,
                           1.781477937,
                          -1.821255978,
                           1.330274429};

    double result;
```

```
if (x<-7.0)
    result = NormalDensity(x)/sqrt(1.+x*x);

else
{
    if (x>7.0)
        result = 1.0 - CumulativeNormal(-x);
    else
    {
        double tmp = 1.0/(1.0+0.2316419*fabs(x));

        result=1-NormalDensity(x)*
                (tmp*(a[0]+tmp*(a[1]+tmp*(a[2]+
                tmp*(a[3]+tmp*a[4]))))));

        if (x<=0.0)
            result=1.0-result;
    }
}
return result;
}
```

Appendix C

A simple array class

C.1 Choosing an array class

To do any serious numerical work, one needs the concept of an array. In C++ this translates into the need for an array class. There are any number of array classes available on the web, and the coder can easily write his own. Indeed most learners of C++ develop their own array class as an exercise sometime early in their study. In fact, the C++ standard library includes an array class which is designed to be optimized for high-speed numerical computation. This template array class is called `valarray` and has different emphases from the standard library's `vector` class. Where `vector` was designed to be efficient for resizing and inserting elements, the assumption in `valarray` is that the array's size will not often vary. Instead `valarray` is designed for speed, and it provides many numerical operations. For a discussion of the `valarray` class see [11] or [29].

There is, however, a downside to using `valarray`: there is no range checking. So if your index goes off the end of the array, you do not get receive a nice error message, but instead you just get garbage or a crash. The reason that range checking is not provided is that it is time consuming, and logically correct code will, by definition, not need it. However, it is an extremely useful debugging tool, and so ideally it should be easy to switch on and off.

Our solution is therefore to provide our own array class which is modelled on `valarray<double>` and provides a subset of its operations. This means that we can use a `typedef` to replace our class with `valarray<double>` when we are confident in our code. This allows us to take advantage of all the optimizations in `valarray`, whilst retaining an easy-to-use class for debugging. Our class provides range-checking if and only if the macro `RANGE_CHECKING` is defined. Thus by changing one project setting, we can quickly shift between safe and fast modes. We also provide the facility to shift between `valarray<double>` and our class via defining the macro `USE_VAL_ARRAY`.

It is important to be careful when choosing an array class. Once a class interface has been chosen, you are stuck with it. You can always gut the insides of your class to make it more efficient but once the class has been used throughout your code, a lot of time has been invested in the interface, and to change it is a difficult proposition.

C.2 The header file

We present the class, called `MJArray` to avoid the likelihood of name clashes, in `Arrays.h`.

Listing C.1 (`Arrays.h`)

```
#ifndef MJARRAYS_H
#define MJARRAYS_H

#ifdef USE_VAL_ARRAY

#include <valarray>

typedef  std::valarray<double> MJArray;

#else    // ifdef USE_VAL_ARRAY

class MJArray
{

public:
    explicit MJArray(unsigned long size=0);
    MJArray(const MJArray& original);

    ~MJArray();

    MJArray& operator=(const MJArray& original);
    MJArray& operator=(const double& val);

    MJArray& operator+=(const MJArray& operand);
    MJArray& operator-=(const MJArray& operand);
    MJArray& operator/=(const MJArray& operand);
    MJArray& operator*=(const MJArray& operand);
```

```
    MJArray& operator+=(const double& operand);
    MJArray& operator-=(const double& operand);
    MJArray& operator/=(const double& operand);
    MJArray& operator*=(const double& operand);

    MJArray apply(double f(double)) const;

    inline double operator[](unsigned long i) const;
    inline double& operator[](unsigned long i);

    inline unsigned long size() const;

    void resize(unsigned long newSize);

    double sum() const;
    double min() const;
    double max() const;

private:
    double* ValuesPtr;
    double* EndPtr;

    unsigned long Size;
    unsigned long Capacity;
};

inline double MJArray::operator[](unsigned long i) const
{
#ifdef RANGE_CHECKING
    if (i >= Size)
    {
        throw("Index out of bounds");
    }
#endif

    return ValuesPtr[i];
}


inline double& MJArray::operator[](unsigned long i)
{
#ifdef  RANGE_CHECKING
```

```
    if (i >= Size)
    {
        throw("Index out of bounds");
    }
#endif

    return ValuesPtr[i];
}

inline unsigned long MJArray::size() const
{
    return Size;
}
#endif // ifdef USE_VAL_ARRAY
#endif // ifndef MJARRAYS_H
```

We have provided the standard class operations: constructor, copy constructor, assignment operator and destructor. We can carry out simple numerical operations, such as addition or multiplication, both by `doubles` which are applied to each element, and by arrays which are applied pointwise.

We have overloaded `operator[]`. It is provided in `const` and non-`const` versions. The former can be used to read elements of `const` arrays, whereas the latter can be used to modify elements of non-`const` arrays. Note that these operators have been inlined. This means that when the compiler encounters the operator, it reproduces the code inside the function instead of setting up a call to the function. This ensures that no extra overhead is caused by going via the class interface. (An alternative would have been to make the data pointer public and use C style array access, but this would have badly violated one of our basic rules that the class data members should be private.)

We also include the `apply` method. This takes in a function object, `f`, and applies it to each element of the array and outputs a new array consisting of the results.

We can take the `size` of our array, and we can resize it. We follow the `valarray` class in not requiring the resize operation to preserve the underlying data.

Finally, we also include the self-explanatory operations `sum`, `min` and `max`. Valarray contains many more operations which could be added to our class as and when necessary.

Our implementation has four data members. We have pointers to express the beginning and end of the array. We also have two different size members. The member `Size` expresses the number of elements currently in the array, whereas the member `Capacity` expresses the amount of memory that has been allocated. So we should always have that Size is less than or equal to Capacity, and that Size is equal to EndPtr minus StartPtr.

C.3 The source code

We present the source code for our array class in Arrays.cpp.

Listing C.2 (Arrays.cpp)

```
#include <Arrays.h>
#include<algorithm>
#include<numeric>

MJArray::MJArray(unsigned long size)
: Size(size), Capacity(size)
{
    if (Size >0)
    {
       ValuesPtr = new double[size];
       EndPtr = ValuesPtr;
       EndPtr += size;
    }
    else
    {
       ValuesPtr=0;
     EndPtr=0;
    }
}


MJArray::MJArray(const MJArray& original)
:
    Size(original.Size), Capacity(original.Size)
{
    if (Size > 0)
    {
       ValuesPtr = new double[Size];

       EndPtr = ValuesPtr;

       EndPtr += Size;

       std::copy(original.ValuesPtr, original.EndPtr, ValuesPtr);
```

```
    }
    else
    {
        ValuesPtr = EndPtr =0;
    }
}


MJArray::~MJArray()
{
    if (ValuesPtr >0)
        delete [] ValuesPtr;
}


MJArray& MJArray::operator=(const MJArray& original)
{
    if (&original == this)
    return *this;

    if (original.Size > Capacity)
    {
        if (Capacity > 0)
            delete [] ValuesPtr;

            ValuesPtr = new double[original.Size];

            Capacity = original.Size;
    }

    Size=original.Size;

    EndPtr = ValuesPtr;
    EndPtr += Size;

    std::copy(original.ValuesPtr, original.EndPtr, ValuesPtr);

    return *this;
}

    void MJArray::resize(unsigned long newSize)
    {
    if (newSize > Capacity)
```

```
    {
        if (Capacity > 0)
            delete [] ValuesPtr;

        ValuesPtr = new double[newSize];

        Capacity = newSize;
    }
    Size = newSize;

    EndPtr = ValuesPtr + Size;
}


MJArray& MJArray::operator+=(const MJArray& operand)
{
#ifdef RANGE_CHECKING
    if ( Size != operand.size())
    {
        throw("to apply += two arrays must be of same size");
    }
#endif

    for (unsigned long i =0; i < Size; i++)
        ValuesPtr[i]+=operand[i];

    return *this;
}
MJArray& MJArray::operator-=(const MJArray& operand)
{
#ifdef RANGE_CHECKING
    if ( Size != operand.size())
    {
        throw("to apply -= two arrays must be of same size");
    }
#endif

    for (unsigned long i =0; i < Size; i++)
        ValuesPtr[i]-=operand[i];

    return *this;
}
```

```
MJArray& MJArray::operator/=(const MJArray& operand)
{
#ifdef RANGE_CHECKING
    if ( Size != operand.size())
    {
        throw("to apply /= two arrays must be of same size");
    }
#endif

    for (unsigned long i =0; i < Size; i++)
        ValuesPtr[i]/=operand[i];

    return *this;
}


MJArray& MJArray::operator*=(const MJArray& operand)
{
#ifdef RANGE_CHECKING
    if ( Size != operand.size())
    {
        throw("to apply *= two arrays must be of same size");
    }
#endif

    for (unsigned long i =0; i < Size; i++)
        ValuesPtr[i]*=operand[i];

    return *this;
}

/////////////////////////////

MJArray& MJArray::operator+=(const double& operand)
{
    for (unsigned long i =0; i < Size; i++)
        ValuesPtr[i]+=operand;

    return *this;
}


MJArray& MJArray::operator-=(const double& operand)
```

```cpp
{
    for (unsigned long i =0; i < Size; i++)
        ValuesPtr[i]-=operand;

    return *this;
}


MJArray& MJArray::operator/=(const double& operand)
{
    for (unsigned long i =0; i < Size; i++)
        ValuesPtr[i]/=operand;

    return *this;
}


MJArray& MJArray::operator*=(const double& operand)
{
    for (unsigned long i =0; i < Size; i++)
    ValuesPtr[i]*=operand;

    return *this;
}


MJArray& MJArray::operator=(const double& val)
{
    for (unsigned long i =0; i < Size; i++)
        ValuesPtr[i]=val;

    return *this;
}

double MJArray::sum() const
{
    return std::accumulate(ValuesPtr,EndPtr,0.0);
}

double MJArray::min() const
{
#ifdef RANGE_CHECKING
    if ( Size==0)
    {
```

```
        throw("cannot take min of empty array");
    }
#endif RANGE_CHECKING
    double* tmp = ValuesPtr;
    double* endTmp = EndPtr;

return *std::min_element(tmp,endTmp);
}

double MJArray::max() const
{
#ifdef RANGE_CHECKING
    if ( Size==0)
    {
        throw("cannot take max of empty array");
    }
#endif RANGE_CHECKING
    double* tmp = ValuesPtr;
    double* endTmp = EndPtr;

    return *std::max_element(tmp,endTmp);
}

MJArray MJArray::apply(double f(double)) const
{
    MJArray result(size());

    std::transform(ValuesPtr,EndPtr,result.ValuesPtr,f);

    return result;
}
```

The code here is quite straightforward. Some points to note: we only reallocate memory when the size becomes greater than the capacity so `operator=` and `resize` check size against capacity. This reduces the number of memory allocations necessary. The data member `EndPtr` is optional in that its value is determined by `ValuesPtr` and `size`. However, having a pointer for the start of the array and the end of the array leaves us very well placed to use the STL algorithms. These generally take in two (or more) iterators which point to the start of the sequence, and to the element after the end of a sequence, which is precisely what `ValuesPtr` and `EndPtr` respectively do.

We therefore use the STL algorithms to perform mundane tasks such as copying, taking the min and taking the max, and soon, rather than writing loops to do them ourselves. As well as saving us coding time, the general principle that one should use pre-defined routines rather than user-defined ones is a good one; pre-defined routines are generally close to optimal and have the advantage that, as part of the standard library, another C++ programmer should recognize and understand them instantly.

Appendix D

The code

D.1 Using the code

The accompanying CD includes source code for all the routines in this book. The code has been been placed in three directories: C/include, C/source and C/mains. Each main program indicate which source files must be included in the same project for the code to link. The include files are included using < > so the directory C/include must be included in the list of places your compiler looks for include files. In Visual C++, the directories for include files can be changed via the menus tools, options, directories.

Makefiles, project files, etc are not included as they are highly compiler dependent.

D.2 Compilers

The code has been tested under three compilers: MingW 2.95, Borland 5.5 and Visual C++ 6.0. The first two of these are available for free so you should have no trouble finding a compiler that the code works for. In addition, MingW is the Windows port of the GNU compiler, gcc, so the code should work with that compiler too. Visual C++ is not free but is popular in the City and the introductory version is not very expensive. In addition, I have strived to use only ANSI/ISO code so the code *should* work under any compiler. In any case, it does not use any cutting-edge language features so if it is not compatible with your compiler, fixing the problems should not be hard.

D.3 License

The code is released under an artistic license. This means that you can do what you like with it, provided that if you redistribute the source code you allow the receiver to do what they like with it too.

Appendix E

Glossary

anti-thetic sampling – a method of improving convergence in Monte Carlo simulations by following each sample by its negative.

class – a user-defined type.

constructor – a member function that has the same name as its class. It provides a way to create objects from the class.

container – a class with the main purpose of holding other objects.

decoration – the act of wrapping a class around another class in such a way that the interface does not change.

encapsulation – the process of representing a concept atomically in terms of a single class.

function – a routine inside a program to which information may be passed and/or returned.

inheritance – defining classes in such a way that they take on the attributes of an existing class plus additional characteristics.

iterator – a class that is similar to a pointer and, in particular, it can be incremented and dereferenced.

member function – a function associated with objects of a particular class.

method – another name for a *member function*.

object – a variable that comes from a class.

pattern – a code design.

pointer – a variable that points to a location in memory.

standard template library – a collection of header files with properties defined by the standard which provide a collection of container classes and algorithms.

STL – shorthand for *standard template library*.

template – a piece of code which is written to work with any class that defines certain chosen methods.

variable – a quantity that is stored within a program and can change in value.

wrapper – a smart pointer class that handles memory allocation and deallocation.

Bibliography

[1] A. Alexandrescu, *Modern C++ Design*, Addison-Wesley, 2001.

[2] M. Baxter & A. Rennie, *Financial Calculus*, Cambridge University Press, 1999.

[3] T. Björk, *Arbitrage Theory in Continuous Time*, Oxford University Press, 1998.

[4] B. Dupire, *Monte Carlo: Methodologies and Applications for Pricing and Risk Management*, Risk Books, 1998.

[5] G. Entsminger, *The Tao of Objects*, Hungry Minds Inc., 1995.

[6] E. Gamma, R. Helm, R. Johnson & J. Vlissides, *Design Patterns: Elements of Reusable Object-Oriented Software*, Addison-Wesley, 1995.

[7] G. Grimmett & D. Stirzaker, *Probability and Random Processes*, second edition, Oxford University Press, 1992.

[8] E. Haug, *The Complete Guide to Option Pricing Formulas*, Irwin Professional, 1997.

[9] J. Hull, *Options, Futures, and Other Derivatives*, fifth edition, Prentice Hall, 2002.

[10] P. Jäckel, *Monte Carlo Methods in Finance*, Wiley, 2002.

[11] N. Josuttis, *The C++ Standard Library*, Addison-Wesley, 1999.

[12] M. S. Joshi, *The Concepts and Practice of Mathematical Finance*, Cambridge University Press, 2003.

[13] I. Karatzas & E. Shreve, *Brownian Motion and Stochastic Calculus*, second edition, Springer-Verlag, 1997

[14] I. Karatzas & E. Shreve, *Methods of Mathematical Finance*, Springer-Verlag, 1998.

[15] A. L. Lewis, *Option Valuation under Stochastic Volatility*, Finance Press, 2001.

[16] S. Meyers, *Effective C++*, second edition, Addison-Wesley, 1997.

[17] S. Meyers, *More Effective C++*, Addison-Wesley, 1995.

[18] S. Meyers, *Effective STL*, Addison-Wesley, 2001.

[19] B. Moro, The full monte, *Risk* **8**(2), 1995, 53–57.

[20] R. Merton, *Continuous-Time Finance*, Blackwell, 1998.

[21] R. Merton, Option pricing when underlying stock returns are discontinuous, *Journal of Financial Economics* **3**, 1976, 125–144.

[22] T. Muldner, *C++ Programming: with Design Patterns Revealed*, Addison-Wesley, 2001.

[23] M. Musiela, M. Rutowski, *Martingale Methods in Financial Modelling*, Springer-Verlag, 1997.

[24] B. Oksendal, *Stochastic Differential Equations*, Springer-Verlag, 1998.

[25] S. K. Park & K. W. Miller, Random number generators: good ones are hard to find, *Comm. ACM* **31**, 1988, 1192–1201.

[26] W. H. Press, S. A. Teutolsky, W. T. Vetterling, B. P. Flannery, *Numerical Recipes in C*, second edition, Cambridge University Press, 1992.

[27] L. C. G. Rogers, Monte Carlo valuation of American options. Preprint, University of Bath, 2001.

[28] A. Shalloway & J. R. Trott, *Design Patterns Explained: a New Perspective on Object-Oriented Design*, Addison-Wesley, 2001.

[29] B. Stroustrup, *The C++ Programming Language*, third edition.

Index